Supplanting America's Railroads

SUPPLANTING
AMERICA'S RAILROADS

The Early Auto Age, 1900–1940

John A. Jakle and
Keith A. Sculle

THE UNIVERSITY OF TENNESSEE PRESS
Knoxville

LIBRARY OF CONGRESS CATALOGING-IN-PUBLICATION DATA

Names: Jakle, John A., author. | Sculle, Keith A., author.
Title: Supplanting America's railroads : the early auto age, 1900-1940 /
John A. Jakle and Keith A. Sculle.
Description: First Edition. | Knoxville : University of Tennessee Press, 2016. |
Includes bibliographical references and index. | Description based on print version record
and CIP data provided by publisher; resource not viewed.
Identifiers: LCCN 2016021910 (print) | LCCN 2016016550 (ebook) |
ISBN 9781621902690 (PDF) | ISBN 9781621902683 (pbk.)
Subjects: LCSH: Railroads--United States—History—20th century. |
Automobile industry and trade—United States—History—20th century.
Classification: LCC HE2751 (print) | LCC HE2751 .J35 2016 (ebook) |
DDC 385.0973/09041—dc23
LC record available at https://lccn.loc.gov/2016021910

To John D. and Irene A. Jakle
and Flory A. and Helene T. Sculle

CONTENTS

ILLUSTRATIONS

PREFACE

We focus herein on the arrival of the motorcar, motor bus, and motor truck, and their impact on America's railroads early in the twentieth century. In 1900 the railroads were supreme. Other forms of transport, such as steamers on the Great Lakes and steamboats on the inland waterways, provided alternative transport, but lacked speed, the ability to operate year round, and, of course, the geographical reach necessary to tie the nation together in an economic sense. Having harnessed steam power for overland travel, the railroad locomotive was king. The nation's roads, for their part, remained principally short-distance feeder lines to rail transport. And then came the automobile. At first the new machines arrived unobtrusively, whether powered by internal combustion engines, electrical motors, or miniature steam engines of their own. The motorcar was initially valued as a vehicle for sport and recreation, and thus as something only the wealthy could afford. At first automobiles seemed little more than a status symbol for the well-to-do. Railroad industry trade journals noted in short news reports the speed and distance records that early motorists achieved. But little did editors suspect that automobility would very quickly become a serious challenge to railroad prosperity. By 1915 the affordable, mass-produced automobile had arrived. And then the motor bus and the motor truck quickly followed. For the first time railroading had a serious competitor, a mode of transport that empowered the individual, through enhanced convenience and flexibility of movement, something that railroading could not provide.

Of course, the electric interurban railroad was another innovation that threatened the railroads, traction cars diverting short-distance passenger traffic from steam passenger trains. But it was a brief role that interurban traction would play. Indeed, it might be said that had motorcars come but a decade sooner, the nation may never have had a traction industry of any size. For a very short period, electric interurban cars, particularly in New England and the Midwest, seemed to promise what automobiles would eventually provide. Built outward from cities, some lines offered long-distance, intercity connection, but most merely serviced small towns, connecting them to nearby cities and thus solidifying urban hinterlands by funneling rural people to city downtowns. But the interurban cars could not offer what the automobile offered: convenience to move more or less at will, flexibility of movement without being confined to a limited right-of-way, and speed of movement (at least before traffic congestion grew). Travel by rail, whether by traction car or by steam train, could not match motoring's apparent freedom of action. Of course, auto ownership came to carry social prestige, not just for the well-to-do but also for the middling classes. It offered the vast majority of Americans a kind of entry into a new modern age not just of personal transport but of personal social enfranchisement, too.

Here we tell the story of how the railroad companies became aware of, how they reacted to, and how they coped with the rise of America's new automobility. The story, however, involves not only the coming of the automobile but also the political circumstances that variously confined the railroads when responding to motoring's challenges. It involves the distrust with which the American people held the railroad industry. A relatively few railroad "barons" had succeeded to industry control by 1900, supported by the nation's financiers. For most localities a kind of monopoly position in transportation had been established. It had brought discriminatory freight rates, for example, and thus economic hardship to far too many people. This had a significant negative effect on farmers in the Midwest, and especially across the Great Plains, giving rise to the Populist political movement, which then morphed into the Progressivism. Congress passed "antitrust" legislation, a principal target being the railroad industry. Thus the Sherman Antitrust Act led to the establishment of the Interstate Commerce Commission with powers to regulate the railroad industry. However, it would not be until the Theodore Roosevelt administration in the opening years of the twentieth century that railroad regulation gained real teeth; this strict regulation came just in time to affect railroad management's reaction to the blossoming auto age.

The railroads responded as best they could. They adopted aspects of the new motoring technology. The internal combustion engine was applied to on-track vehicles used in railroad maintenance and repair. Then the self-propelled rail motor car was introduced as a means of replacing more expensive steam trains, particularly on less profitable branch line runs. Rail motor cars wrought many of the same advantages promised by the electric interurban cars: more frequent and better passenger service on branch lines, for example. Originally, automobiles themselves were actually seen as benefiting railroading. Taxicabs, rather than horse-drawn cabs, and then private autos brought passengers to and from railroad stations. When motor trucks picked up and delivered package freight,

it extended a railroad's reach to customers so much faster and much more efficiently than teamsters with horse-drawn wagons. And then the railroads began
to operate their own trucks, many railroad companies establishing trucking subsidiaries to do so. Following the lead of electric interurban and city streetcar operators, the railroads began to operate buses, and railroad bus subsidiaries came
to the fore. Indeed, railroad bus subsidiaries were substantially responsible for
organizing the nation's bus industry, including today's Greyhound Lines, Inc.,
and Trailways Transportation Systems, Inc.

At first the railroads welcomed rural highway improvement. How better to
speed farm products, for example, to railroad sidings? And how better to extend
the trade reach of a railroad town as large railroad systems competed with one
another for territorial dominance across broad regions? Many railroad companies played an active role in the nation's early "Good Roads Movement," which
led to rural highway improvements first at the state level and then at the federal
level. As automobile sales soared, the railroads benefited by carrying raw materials to auto plants, and new cars to market, as well as crude oil and finished
gasoline to and from refineries in tank cars. But motoring and railroading also
came into conflict, indeed physical conflict—especially at railroad crossings,
the one place where the two technologies physically competed. Through it all
the railroad industry persevered but against increasing odds. State and federal
governments subsidized highway construction to sustain motoring in its many
forms. The railroads, on the other hand, continued to build and maintain their
own rights-of-way. And they paid taxes on those rights-of-way. Commercial
motor carriers—the nation's bus and truck lines—either did not pay taxes for
road use or did so at relatively low costs. Of course, the federal government
also subsidized the barge lines on the nation's inland waterways (for example,
by building locks and dams). And it subsidized the nation's new airlines (for example, with government-funded airports and cross-country air beacons). Not
until the mid-1930s were serious attempts made to "level the economic playing
field"—something the railroads, through decades of strenuous lobbying, had
long sought.

Our period of concern is limited to the first four decades of the twentieth
century—the early auto age, as we call it. The first motor vehicles appeared on
America's city streets and rural roads around 1900, our starting date. Through
the 1910s and 1920s, as motoring took hold of the nation's imagination, railroading slowly declined as an economic enterprise, a decline that notably accelerated

once the Great Depression hit in the 1930s. World War II quickly reignited railroad fortunes. In 1940 the nation was poised on the threshold of war, having yet to fully recover from the deepest economic depression it would ever experience. But the bombing of Pearl Harbor and America's entry into what was already a worldwide conflict fully reversed the nation's economic doldrums, enabling the railroads to substantially recover levels of freight business previously lost. Significantly, gasoline rationing severely constrained the private use of automobiles, and thus railroad passenger travel rebounded. It is with the outbreak of World War II, therefore, that we close our investigation. After the war, of course, the motorcar, along with the motor bus, but especially the motor truck, assumed their growing predominance. But for the railroads the economic turnaround proved but a temporary respite. Only in the 1980s would Congress finally relent on oppressive railroad regulation, enabling another railroad revival. Automobility's impact on railroading during and after World War II, however, is a story to be told elsewhere.

In the chapters that follow we not only seek to tell the important story of how the railroads coped with the nation's new automobility, but we do so through the eyes and ears of those editing and writing for industry trade journals. We are especially reliant on *Railway Age,* the railroad industry's principal newsmagazine. *Railway Age* matured as a news service through numerous mergers, consolidating with other magazines bearing such names as the *Railroad Gazette, American Railroad Journal, Railroad & Engineering Journal,* and *Railway Locomotives & Cars.* It assumed its present name in 1910. But we have also relied secondarily on the electric railroad, bus, and trucking journals, especially *Electric Railway Journal* (which began publication in 1908), *Motor Truck News* (which began in 1911), and *Bus Transportation* (which began in 1922). Concern with trade-journal editorials, news briefs, and articles enables us, we believe, to better establish for readers the tone of the times. What was it that editors and journalists thought industry readers ought to know? What would they have them advocate? In other words, what were the prevailing arguments put forth regarding motorcars, buses, and trucks? And for what purpose? And with what apparent result?

In previous books we have sought to cover in its many dimensions the story of how automobility came to impact the American experience, especially through changes in the nation's built environments. We began by focusing on the rise of roadside America with its new commercial fixtures such as gas stations, motels, and roadside restaurants. We went on to consider various other automobile-related topics: highway improvement, motoring and night lighting, outdoor advertising, the garage as an American institution, and the rise of the lowly parking lot as a new kind of land use. We, of course, were not alone in authoring such books, as we reviewed in our recent book, *Remembering Roadside America: Preserving the Recent Past as Landscape and Place,* also published by the University of Tennessee Press. In that book we emphasized how very little of the nation's early-twentieth-century automobile-impacted landscape survives, so ever-changing has automobility's impact been on the American scene. And so also has automobility brought constant change to railroading and its distinctive landscapes and places. That is the theme we explore here.

Chapter 1

AMERICA'S RAILROADS EARLY
IN THE TWENTIETH CENTURY

American railroads fully dominated freight and passenger transport in the United States in the early twentieth century. The largest of the nation's railroads, including the Pennsylvania Railroad and the Vanderbilt-controlled New York Central, numbered among the world's largest corporations. Their grand passenger terminals in the nation's largest cities spoke of economic power and prestige and, for many Americans, corporate greed as well (fig. 1.1). Nonetheless, railroading's premier standing had begun to decline well before World War I, with the coming of the motorcar, motor truck, and motor bus a contributing factor. Although our main interest in writing this book concerns motoring's impact on railroading, we first need to consider how other factors, particularly government regulation, played a role in that decline. The railroads paid little attention when the first automobiles arrived in the 1890s. At first the automobile seemed little more than a passing fad, then as something that might benefit and not hurt the railroads, but then in the 1920s as something threatening. Until then there were other, more serious worries, specifically restrictions imposed by state and federal governments. Government regulation impacted railroad ownership, financing, railroad operations, and service, thus creating a political and economic environment substantially supportive of competing transportation modes, especially,

FIGURE 1.1. Entrance to Pennsylvania Station, New York City. Front cover of Pennsylvania Railroad timetable, April 30, 1933.

as it would turn out, highway development, but also transport on the nation's inland waterways and in the air.

Railroading in the United States originated in the 1820s with horse- and mule-drawn cars rolling on twenty-three miles of rail by the end of the decade. Through the 1830s, with an increase in track mileage and, more important, the introduction of steam locomotives, most railroads served mainly as feeder lines to water transport. By the 1850s line expansion outward from eastern seaport cities (especially New York, Philadelphia, and Baltimore), coupled variously with midwestern connecting lines (many of them built as subsidiaries, provided connections to cities like Chicago and St. Louis), marked the beginning of trunk-line railroading, linking the Midwest's economy firmly with that of the Northeast. By 1841 some sixty-five hundred miles of railroad were operation, some nine thousand miles by 1850, and over thirty thousand by 1860.[1] Railroading in the Southeast lagged behind. Long-distance, or "trunk line," connections there remained quite undeveloped before the Civil War. During the war railroads played an important role in northern military strategy, greatly facilitating the Union's transport of men and matériel to armies in the field. After the Civil War the railroads contributed substantially to the North's industrialization, not only by connecting distant points, but also through purchase of iron and then steel rail, among other commodities.

Initially, state governments, as well as the federal government, offered important financial support. In Illinois, for example, the state launched a network of north-south and east-west railroads. The ambitious project abruptly ended with the economic depression of 1837, bankrupting not only the railroads but the state. Illinois was not alone in launching railroad building through generous monetary subsidies, but also through outright gifts of public land, and allowing eminent-domain privileges so that railroads might readily take privately owned land for rights-of-way. In most instances, the railroads were privately owned companies and not governmental entities. Importantly, state governments issued corporate charters, often with monopoly implications regarding service to specific cities. In Illinois, the U.S. Congress ceded some 2.5 million acres of unsold public domain to the Illinois Central Railroad, land relinquished in a checkerboard pattern of alternating one-mile-square sections along its proposed main line. Subsequent sale of land enabled the railroad to complete its seven-hundred-mile main line (creating the world's longest railroad at the time), south to north from Cairo, where the Ohio and Mississippi Rivers met, to Galena, a lead-mining center and then one of the state's largest cities. In lieu of taxes the railroad annually paid the state 7 percent of its gross earnings on that line. Additionally, the railroad was required to carry the U.S. Mail at 20 percent of regular rates and U.S. troops at 50 percent. The Illinois Central was a key player in the Union army's successful control of the lower Mississippi River during the Civil War, a branch line having been completed, angling from the town of Centralia in southern Illinois northeast to Chicago.

In total, some fifty railroads nationwide eventually received federal land grants amounting to some 155 million acres. After the Civil War several transcontinental railroads were so subsidized, including the Union Pacific and its link westward, the Central Pacific, as well as the Northern Pacific and the Atchison, Topeka & Santa Fe.[2] Public largess enabled these and various other railroads to extend into unpopulated territories, thus encouraging a faster-moving frontier of settlement. Railroad building during the 1870s and 1880s across the trans-Mississippi West did not follow population movement so much as precede it. Unfortunately, government subsidies too frequently led to harmful speculation, including stock manipulation and other financial abuses, all of which severely tarnished the railroad industry, leading to public demand for railroad regulation. Indeed, the railroads did abuse their rate-making privileges too often, favoring one or another category of shipper over others, or various localities over another. It was from such abuse that the Populist political movement evolved in the 1880s.

Railroad mileage (defined in terms of single-track rights-of-way) reached nearly 53,000 miles in 1870, 93,000 in 1880, 159,000 in 1890, and 193,000 in 1900. However, when second, third, and fourth main-line tracks, along with trackage in yards, were counted, the total jumped to some 259,000 miles.[3] In 1900 America's railroads operated some 38,000 locomotives, 1.3 million freight cars, and 35,000 passenger cars. They employed some 1.1 million workers with compensation topping over $500 million annually.[4] Along with their families, railroad workers accounted for some 8 percent of the nation's population.[5] Railroading was at its peak as an important contributor to the nation's economy. Railroad assets made up approximately $14.5 billion of the nation's estimated $90 billion total wealth. Only agriculture exceeded the railroad industry in the amount of capital invested and also the amount of annual business conducted.[6] Railroads consumed more manufactured products than any other economic sector. For example, of the 18 million tons of iron and steel produced in 1907 in the United States, the railroads bought over half. They bought a quarter of all the timber and lumber sold.[7]

RAILROAD REGULATION BEFORE WORLD WAR I

Beginning in the 1870s, public hostility toward the railroads grew rapidly. State governments as well as the federal government assumed a very different stance toward the railroads. Instead of offering incentives to lay track and assuming rather laissez-faire stances regarding railroad operations as before, those governments brought degrees of direct regulation to the fore. True public hostility was fostered by a host of railroad abuses, mainly the preferential rates and services offered to large shippers over small ones, as well as the industry's predilection for long-distance haulage over short-distance. Certain cities and whole regions were privileged through establishment of differential freight rates. In general, the nation's railroads charged as much as traffic would bear. Along any given line a railroad company enjoyed a kind of monopoly except where the lines of a competitor closely paralleled or crossed. Lack of competition in a locality not only enabled rate abuse but encouraged it, with local customers having no choice but to pay up.

National agitation for freight-rate relief crystalized first among grain farmers of the Great Plains, fostering the so-called Granger Movement. Thus did states like Illinois, Iowa, Minnesota, and Wisconsin lead the way. Unreasonable rates and unjust discriminatory practices were prohibited there by statute. Schedules of maximum rates to be observed by the railroads were set by legislative enact-ment, and state commissions were organized for their enforcement.[8] Previous state oversight nationwide, such as it was, had focused more on inspection of physical plant, collection of operating statistics, and investigation of suspected charter violations. Then in 1887 Congress established the Interstate Commerce Commission (ICC) to oversee the railroads nationwide. Following challenges in the courts, the Supreme Court ruled such oversight to be constitutional in 1898. Critically, the majority report read: "A railroad is a public highway, and none the less so because constructed and maintained through the agency of a corporation deriving its existence and powers from the State. Such a corporation was created for public purposes. It performs a function for the State."[9]

Into the first decade of the twentieth century, railroad regulation sought to (1) correct excessive rate inequalities by establishing "just and reasonable" rates, (2) assure sustained, efficient, and safe operation of the railroads, (3) assure proper working conditions for employees, and (4) encourage investment in railroad securities. Central was the five-member (later increased to eleven-member) In-terstate Commerce Commission, which had restricted regulatory oversight but had no direct power to actually enforce decisions rendered; rather, litigation in the federal courts was necessary to do so. The railroads were required to periodi-cally file detailed reports regarding operations, labor practices, and financing, following standardized accounting procedures. Congress amended and enlarged the Interstate Commerce Act repeatedly: for example, with the Safety Appliance Act or Cullum Act of 1893, which gave the ICC jurisdiction over railroad safety, removing it from the states; the Hepburn Act of 1906, which broadened ICC oversight responsibilities to include oil pipelines, express companies, sleeping car companies, but also the setting of "maximum railroad rates," allowing the railroads to actually reduce rates when and where appropriate; the Mann-Elkins Act of 1910, which created a commerce court, making the ICC the final arbitrator of railroad rate decisions and thus bringing to the ICC real power of rate enforcement; the Railway Valuation Act of 1913, intended to encourage freight rates based on the real value of railroad infrastructure rather than on what critics contended to be the "watered-down" value of railroad stock; and the Adamson Act of 1916, which established an eight-hour workday for interstate railroad workers. Various new boards were created, including the National Mediation Board (1934), charged with minimizing railroad work stoppages, including strikes, and the Railroad Retirement Board (1935), charged with administering new retirement benefits for railroad workers. It is generally conceded that before 1916 the federal government had little impact on how the railroads operated and how they were financed. Indeed, it was through railroad financing that much federal oversight was effectively circumvented.

Perhaps, the Sherman Antitrust Act was more restrictive as it assigned potentially monopolistic business practices, including those of the nation's railroads. Thus were the railroads prevented from sustaining their "pooling" arrangements once the Supreme Court declared them illegal.[10] Railroads, being overbuilt in most parts of the nation, had joined in traffic- and revenue-sharing combines, setting rates and sharing profits, especially through long-distance freight and passenger service. To evade the court's decision, there followed a wave of railroad mergers accomplished through various means: for example, through lease or outright purchase of one railroad by another or, more commonly, through purchase of a controlling interest in another via a holding company. In consequence, ownership of the nation's railroads became highly concentrated in the hands of only a few investor syndicates. Rather than restrict potentially monopolistic practices, the court's anti-pooling decision had quite the opposite effect.

At its core, a holding company was a voting trust that purchased and held shares of capital stock in one or more operating corporations, thus to exercise "all rights, powers and privileges of ownership including the authority to vote the shares and receive the dividends."[11] Some railroad trusts, once they acquired a railroad, allowed it to keep a separate identity as well as a separate management, but usually with cooperative agreements regarding traffic exchange. With other trusts, operating companies, once they had been acquired, were fully integrated. Increased consolidation tended to follow periods of economic depression when weaker railroads, having gone into receivership, could be purchased at rock-bottom prices by stronger roads. Of course, many bankruptcies, especially during economic hard times, were fostered by investor syndicates overreaching through highly leveraged stock purchases. In that regard, the nation's banks came to the fore as the commanding means of rescue, with major investment houses such as J. P. Morgan and Company and Kuhn, Loeb and Company leading the way. By 1906 eight major railroad trusts had been created, not necessarily in direct response to ICC oversight, but with full implications for eluding that oversight. The eight railroad groupings included:

> The Vanderbilt System, built around the New York Central and Hudson River Railroads (and its many subsidiaries across upstate New York and the Midwest), along with the Nickel Plate (the New York, Chicago & St. Louis), the Lackawanna (the Delaware, Lackawanna & Western), and Chicago & North Western, for a total of 22,578 miles of right-of-way.

> The Pennsylvania System, built around the Pennsylvania Railroad, with the Baltimore & Ohio, Chesapeake & Ohio, and Norfolk & Western in addition, for a total of 20,370 miles.

> One Morgan System, which included, among other lines, the Erie, the Mobile & Ohio, the Lehigh Valley, and the Southern Railway, for a total of 17,810 miles.

> A second Morgan System, which included the Atlantic Coast Line and Louisville & Nashville and their various subsidiaries, for a total of 11,202 miles.

> The Gould System, built around the Wabash and the Missouri Pacific, with the St. Louis Southwestern, Texas & Pacific, and Denver & Rio Grande in addition, for a total of 16,902 miles.

> The Harriman System, built around the Union Pacific and its subsidiaries, with the Southern Pacific and Illinois Central in addition, for a total of 19,182 miles.

> The Hill System, including the Great Northern, the Northern Pacific, and the Chicago, Burlington & Quincy, for a total of 21,303 miles.

Mapped are the main lines of the Vanderbilt System, which stretched from the Atlantic to the Rocky Mountains (fig. 1.2).

Thus, some two-thirds of the nation's railroad mileage was under the control of just seven railroad ownerships.[12] However, the ability to merge or

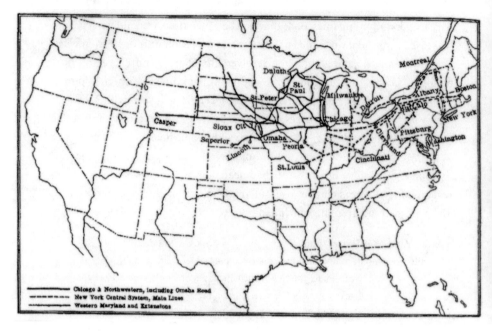

FIGURE 1.2. Railroads controlled by the Vanderbilt Trust, 1920. C. M. Keys, "The Shifting Railroad Control," *World's Work*, 20 (June 1910): 1304.

consolidate operating companies was diminished in 1904 when the Supreme Court ruled against North Securities, the holding company James J. Hill had used to merge his three railroad holdings.[13] Hill was forced to break them up. Federal court decisions hinged mainly on interpretations of the Sherman Antitrust Act of 1890, but also on the Clayton Act of 1914, which strengthened the former by making illegal any corporate merger or, for that matter, any cooperative venture that would lessen competition within an industry. Thus the railroads were prevented from increasing operating efficiencies through the full integration of owned companies. Degrees of operating coordination had been sustained through interlocking directorships and individual managers holding similar positions at related companies. But that the Clayton Act eliminated.

Many of the nation's stronger railroads continued to modernize, and new lines continued to be built. Over three decades leading up to 1912, the industry added each year on average some six thousand miles of new railroad right-of-way, although thereafter new construction fell off sharply.[14] As the editors of the *World's Work* commented (but not without a bit of hyperbole), "Side by side with the centralizing of railroad power ran a related process of intensive growth with the great systems. Instead of branches, the huge trunk-lines built sidings, bought cars, put new engines into service, split mountains open, drained rivers, pierced the depths with tunnels, urged the investors onward toward a climax in the weight of engines, standardized equipment—did a thousand things that all pointed in the direction of efficiency and intensive growth."[15]

The federal government was not alone in regulating the railroads. State government boards and commissions continued their work, but focused, of course, on intrastate (as opposed to interstate) commerce. In assessing the regulatory impact of the states, journalist French Strother noted, "Within three years of 1907, twenty-five states enacted car-service laws, twenty-three regulated train service and connections, twenty-two fixed maximum passenger rates, nine enacted maximum freight rates, thirty-six regulated the general corporate affairs of common carriers." In total, thirty-three states had enacted 334 laws regulating railroads within their jurisdictions.[16] The ICC, it should be acknowledged, had authority over railroad affairs within states but only to the extent that those affairs impacted the interstate movement of freight and passengers. The ICC could and did countermand state regulations.

TAXATION

Besides regulation, the railroads found their tax liabilities rapidly increasing. In the nineteenth century, many states assisted the railroads by allowing tax rebates and/or setting tax rates especially low. With growing public hostility toward railroad rate making and with increased distrust of the large railroad trusts generally, these policies were largely reversed, and railroad infrastructure was deliberately taxed as much if not more than that of other industries. Property taxes represented the largest tax liability. Before 1880 local tax officials assessed corporate real estate, set tax rates, and submitted tax bills. After 1880 state boards in most states took control, fixing the value of each railroad property statewide, setting the property tax rate, and then distributing tax revenues from locality to locality according to the size of a railroad's local operations. Some states taxed the holders of railroad stocks and bonds. Some imposed a tax on gross receipts. Some imposed a so-called "franchise tax" based on formulas combining assessments of receipts, profits, dividends, value of capital stock, amount of bonds, and so on.

Locomotive vs. Truck

THE LOCOMOTIVE SAYS:	THE TRUCK SAYS:
1—"My Company owns the road over which I run.	1—"My company owns no highways.
2—"It pays heavy taxes to the state and to every city and county in which it has tracks.	2—"The public is taxed to build my road, and taxed again to keep it in repair.
3—"It pays a large part of the school tax and thus helps educate your children.	3—"We contribute little or nothing toward the education of your children.
4—"It employs many thousand section men, and at its own expense keeps its road in condition for use.	4—"We pay only a small part of the taxes for road."
"I PAY MY WAY."	THINK IT OVER!

FIGURE 1.3. Advertisement from *Official Railroad Rate Book*, Champaign-Urbana (Illinois) (N.p., 1939–40), 2.

In 1906 the total amount of property taxes paid by American railroads totaled some $75 million. In three states—Illinois, New York, and Pennsylvania—payments exceeded $5 million. The heaviest taxes per mile of road were paid in Massachusetts, which received $1,683 per mile, with Connecticut second at $1,220 per mile. Although the total taxes paid in Illinois amounted to $5,364,330, the sum per mile was only $453. Taxes paid on the value of stocks and bonds held by the railroads, and the earnings received from them, totaled some $8 million. However, Emory R. Johnson, a professor of transportation and commerce at the University of Pennsylvania, concluded that railroad taxation in 1906 was not excessive, certainly in regard to the taxing of stocks and bonds. In subsequent years, as railroad revenues declined, the railroads would argue quite differently.[17] However, it was not so much what the railroads paid in taxes as how much other modes of transport did not. By 1920 the states and the federal government had come to invest heavily in highways and inland waterways and, by 1930, in airports. Truckers' increasing use of the highways would prove especially vexing (fig. 1.3.)

WORLD WAR I

By 1915 it was clear that the railroad capacity was not keeping up with demand. Over time stabilized railroad rates had meant reduced revenues, and reduced revenues had meant reduced investments in rolling stock and in the expansion of main lines from one to two and three tracks where warranted. Serious congestion developed particularly in the Northeast. Seasonally, especially in the agricultural states of the Midwest, grain-car shortages occurred year after year. Pooling was no longer possible to encourage the efficient handoff of traffic from one railroad to another, nor was consolidation possible to fully integrate and thus create efficient long-distance operations under single management. With war preparations underway as the nation's involvement in World War I became increasingly likely, the federal government acted. The Army Appropriation Act of 1916 empowered the president to nationalize the nation's railroads during wartime or other national emergencies. Acting on this authority, Woodrow Wilson moved to do so in December 1917, with Congress passing the Federal Control Act that called for a director general of railroads. Initially, the idea was for railroad management to cooperate voluntarily, and a railroad war board was created to encourage that. In 1917 the railroads did increase their handling of freight by some 10 percent over the year before despite the loss of some seventy thousand employees to the military.[18] The placing of larger freight cars in service, the use of longer trains with newer more powerful locomotives, and the speeding up of train loading and unloading at terminals all helped.

Car shortages, nonetheless, prevailed. Rail yards remained congested. The flow of war matériel destined for Europe slowed, so much so that the federal government began to subsidize truck convoys between factories in the Midwest and docksides at eastern seaports. Created, therefore, was the United States Railroad Administration, charged with direct and complete control of railroad activity nationwide. Many ICC-imposed restrictions on railroad operations were lifted, with pooling arrangements, most importantly, freed from restraint. Pooling sped the movement of empty cars to locations where needed. The railroads, however, were given financial assurances. They were guaranteed, for example, a standard return of 6 percent for their stock and bond holders, based on the earnings of the 1914–1917 period. The government promised to maintain and even expand the railroads' physical plants, creating a revolving fund to do so. In 1918 the government loaned the railroads some $180 million, made direct investments amounting to some $103 million, spent $445 million for "additions and betterments," and allocated $118 million for increased materials and supplies.[19] In 1918 the Railroad Administration ordered nearly two thousand new locomotives and one hundred thousand new freight cars.[20] After the war, nonetheless, the railroads charged the government with neglect: railroad rights-of-way and terminals, it was claimed, were rundown and in poor condition.

THE TRANSPORTATION ACT OF 1920

The Transportation Act of 1920 (the so-called Esch-Cummins Act) returned the railroads to private ownership but only after a period of debate in Congress regarding continued federal control if not ownership. But it was generally agreed in Congress that the railroads, if nationalized, would be inefficient, a belief based less on wartime performance than on ideological grounds—the belief that railroad control by private capital was fundamentally preferable to control by government bureaucracy. Under federal management and despite federal spending, the railroads had been denied inflation-matching rate increases, which in fact had deferred essential maintenance. The industry's ratio of operating expenses to operating revenue had deteriorated from a respectable 76 percent to an unacceptable 95 percent. Government generosity to the various railroad unions had been lavish. Prior to takeover, the employee share of each revenue dollar had been 40.3 cents; in 1920, however, it was 55.4 cents.[21] Once the railroads were privatized, the Interstate Commerce Commission assumed its prewar regulatory role. And many former regulations came back into force.

The word "just" appeared forty-eight times in the Interstate Commerce Act, and the word "reasonable" sixty-six times. Thus the ICC was charged with seeing that rates, fares and other charges, and operational practices were just and reasonable, with emphasis on protecting shippers. But now the ICC was also charged with protecting the railroads and therefore ensuring the nation an adequate transportation system. The 1920 act allowed the ICC to set minimum as well as maximum rates. The ICC was also charged with preparing a plan whereby America's larger "Class I" railroads might be consolidated, a task delegated to

William Z. Ripley, an economist at Harvard University. The Ripley plan proposed twenty-one regional railroads. Consolidation was not to be imposed; rather, the plan merely suggested how individual railroads might volunteer to merge. The main idea was that, in principle, economically stronger roads would absorb the weaker. To facilitate lease and financial-control transactions, the 1920 act substantially diluted the Clayton Act's absolute rejection of interlocking directorates—one person serving as an officer or director of more than one company. ICC decisions regarding rates and other matters were to be made fully in the "public interest," a concept left largely undefined. Eventually, however, the Transportation Act of 1940 would repudiate the Ripley plan, causing the consolidation idea to be abandoned after years of inaction.

Some three hundred applications for lease or financial control were brought to the ICC through the 1920s and 1930s. Most of them concerned Class I railroads absorbing short lines. And most were thus approved. But efforts to merge Class I railroads were largely rejected, mostly on the grounds that proposed financing would not equally serve every interest. Investors in the weaker roads needed to be protected. Yet, healthy railroads refused to accept that weaker railroads should be valued on par. Among the proposals made was one brought by the Van Sweringen brothers, Oris and Mantris. Real estate developers in Ohio, they had bought an electric railroad to connect suburban Cleveland's Shaker Heights with a new terminal and office building in downtown Cleveland. Then they bought the New York, Chicago & St. Louis Railroad (the Nickel Plate), merging it with several regional lines in the Midwest to form a new midwestern system, one of the few such mergers approved by the ICC. But they were turned down in seeking an additional merger with the Chesapeake & Ohio (and its various subsidiaries), as well as with the Erie, thus to form a new trunk line competitive with the New York Central, the Pennsylvania, and the Baltimore & Ohio. The Van Sweringens had become masters of the leveraged buyout, whereby companies were purchased, their assets sold off, and subsequent revenues diverted to cover borrowing costs.

Their second proposal hinged on their controlling various holding companies (including the Alleghany Company, the Chesapeake Corporation, the Clover Leaf Company, the General Securities Corporation, the Nickel Plate Securities Corporation, the Vaness Company, the Virginia Transportation Company, and the Western Company). Their plan, the ICC held, involved "an utter lack of independent and impartial representation of stockholders." If approved, the brothers would fully control the new system, owning only one-third of voting stock. Previously, large railroad systems had been put together by various means: leases, traffic agreements, trackage agreements, majority share interest, joint control and guarantee of outstanding or newly incurred debt, and so on. Holding companies, of course, involved multiple layers of control. But as the 1930s wore on, only one kind of unification seemed to meet ICC standards: corporate consolidation into a single company for ownership, management, and operational purposes. But the Transportation Act of 1940 repudiated the Ripley plan, making such consolidation improbable. "There is an apparently growing opinion," wrote Pierce H. Fulton in the *Magazine of Wall Street*, "that consolidation would be too unwieldy to be efficient, that it is politically undesirable, undemocratic, and virtually tending toward government confiscation."[22] However, more than a little consolidation had already been accomplished—if not in managing railroads, then in owning them. In 1931 Class I railroads operated on some 241,000 miles of right-of-way. Sixteen railroads and forty-six subsidiaries controlled approximately 146,000 of those miles or about three-fifths of the total. Some 21,000 miles were under the ownership of a single individual or family trust.[23]

What the Transportation Act of 1920 did not do, despite its all-embracing name, was to link railroad regulation with the similar regulation of other transportation modes. Motor trucks and buses, for example, escaped federal oversight, as we discuss in detail in subsequent chapters. Important here is the fact that the ICC was not charged with considering a railroad as potentially part of a larger transportation system involving multiple modes. Congress focused on maintaining levels of competition between the railroads, thus to reduce monopolistic practices in rate setting and so forth. Multimode transportation companies involving railroads and highway transport most specifically (but even water, pipeline, and air transport as well) were little considered as a means of not only engendering competition, but also making transportation more efficient. A National Transportation Conference, sponsored by the United States Chamber of Commerce in 1924, did call for ICC oversight of motor carriers, thus to insure that trucks and buses paid a "fair share" of the burden of public expenditure for highways. Better coordination between rail and highway transport was advocated, but even there each transport mode was viewed in isolation. The Transportation Act of 1920 was exceedingly conservative in that renewed federal oversight of the railways did not take into consideration the nation's new (or renewed) transport technologies, especially the challenges wrought by automobility. At the time when motorcars, trucks, and buses came to the fore, the railroads were being regulated for past sins instead of being encouraged toward an innovative future. At the very time that federal government began to subsidize highway and waterway improvements, and the creation of airports as well, the railroads found themselves regulated as if such competing transportation modes did not exist.

DECLINING RAILROAD FORTUNES

With the increased powers of the ICC, especially the commission's stabilization of railroad rates to favor shippers, the value of railroad stocks and bonds declined. Limitations placed on railroad rates in 1906 depreciated the value of railroad securities and helped cause the Panic of 1907. The overall value of railroad securities then declined by half over the ensuing decade, although, admittedly, government regulation was not the sole reason. Yet regulation's incremental impact was clearly long-term, greatly enhanced with the Transportation Act of 1920, which set 5 to 6 percent of capital as the basis for establishing fair annual railroad profits, with railroad rates to be adjusted accordingly. Six percent was far below what other industries were earning—for example, the new automobile industry. Before World War I, railroad stocks were far more popular investments than industrial stocks. In 1901 a total of some 188 million railroad shares changed hands on the New York Stock Exchange, but only 38 million in 1923. In 1901 some 70 percent of all stocks traded were railroad issues, but only 16 percent twenty-two years later. In 1902 the average price for a railroad share was about $130, with industrial shares selling for about $70. During World War I, when orders for industrial commodities vastly increased, industrial stocks soared in price. They retained their value even with postwar dollar inflation. Railroad shares, on the other hand, lost about half their value in the 1920s.[24] Railroads came to be dependent primarily on bonds to fund new infrastructure and rolling stock. However, the cost of bond issues increased, given their declining popularity. The railroads turned from bonds with fixed interest to so-called "adjustment" and "income" bonds, which did not pay interest unless revenues warranted, being designed to keep railroad companies out of receivership. Railroad bonds, once investor preferred, gave way to other new tax-exempt instruments such as municipal bonds. Such changes left the railroads substantially undernourished.

In 1914 James J. Hill of the Great Northern mused on the difficulty that railroads were then having raising new capital: "With pressure upon them from all sides for more wages, more taxes, more facilities, more kinds of costly service and more money to pay salaries of State and national agencies charged with the never ending task of investigation and inquisition, they cannot even maintain the unsatisfactory rate of earnings of their recent past. This is not the argument of an advocate for a cause; it is the conclusion drawn from official facts."[25] Also in 1914 Ivy L. Lee, executive assistant for the Pennsylvania Railroad, summarized the difficulties his company faced: (1) floods ($12 million of damage in 1913 alone), (2) increased wages (compensation paid out had increased 37 percent over 1911), (3) taxes (tax bills increased by 111 percent over ten years), and (4) demands of new government legislation (state laws, including those covering grade-crossing improvements, were up some $6 million). And then there was confusing regulation. As Lee wrote, "We are compelled by the Sherman Act to compete; and yet, under the Hepburn Law, in substance forbidden to compete with other railroads. The Hepburn Law, by requiring us to file all rates thirty days before they become effective, practically insures that the rates over railroads between two points shall be the same."[26] Lee was further quoted in the *World's Work*: "Lack of railroad credit is due to the present chaos of regulation by forty-three state commissions and the Federal Interstate Commerce Commission. These commissions prescribe the kinds of cars, number of train crew, hours of labor, style of cars, kind of headlights, brakes, signals, etc., and the regulations of the different commission are often, if not usually, contradictory. . . . the commissions also fix the freight and passenger rates, which govern the income."[27]

Railroad officials could not help but envy what was happening in other industries. L. E. Johnson, president of the Norfolk & Western Railway, praised what he saw in the auto industry:

> The growth and magnitude of the automobile industry is startling and instructive, and the growth of the electric industry alone rivals it. Together they are modifying our social life in a bewildering manner, and have an economic importance of the first rank. The value of automobiles sold in the last fiscal year was $625,000,000 and exports reached $100,000,000. The industry consumed 13,000,000 barrels of gasoline and 20,000,000 gallons of lubricants. Eight million tires were used, and there were 51,000 workmen employed in a single State, while there were 448 factories in 34 states.

Comparing motoring with railroading, he continued:

> It is said that motor vehicles traveled ten billion miles. If there were not more than three passengers in each vehicle, that is thirty billion passenger miles. The

steam railways report that in the same year they carried over one billion passengers an average of 34 miles for each journey, or a passenger mileage of 34 billion.

And yet it was the automobile that was being cherished and pampered, if only through the subsidizing of public roads:

> Tens of thousands of miles of good roads have been built that never would have been built but for the automobile, and the nation soon will be relieved of the curse and the waste that bad roads entail.[28]

But where were the regulations concerning motorcars, trucks, and buses? Railroads, being regulated, had been placed in a losing position vis-à-vis motor vehicles and their lack of regulation. In 1909 alone, 664 railroad laws were enacted in forty-one states. Between 1912 and 1915 upwards of 4,000 bills affecting railroads were introduced both in Congress and various state legislatures, 440 of which became law. And then there was taxation. Taxes per mile of railroad line increased from $290 in 1904 to $560 in 1915.[29] What were motorists paying?

The health of America's railroads continued its downward trend during the 1920s and 1930s, undermined, of course, by the nation's new highways that not only invited more and more private motorists to use but also encouraged the proliferation of commercial trucks and buses. In 1920 the railroads carried almost everything long distances, freight as well as passengers: 2.5 billion tons of freight and 1.2 billion passengers. But during the 1920s government at various levels spent some $1.7 billion on inland waterways and harbors and improved and surfaced nearly a million miles of rural highway. Between 1920 and 1929 the number of registered motor vehicles in the United States nearly tripled from some 10 million to nearly 27 million.[30] Still very much the backbone of the nation's transport system in 1929, the railroads retained some 75 percent of the nation's freight haulage, but ten years later it was down to 69 percent. Trucks, on the other hand, increased their share from just over 3 percent to 10 percent. (The share moved by barge on the nation's inland waterways jumped from 17 percent to 19 percent, and in pipelines from approximately 4 to 2 percent.)[31]

THE EMERGENCY RAILROAD TRANSPORTATION ACT OF 1933

It was the stock market crash of 1929 and the Great Depression that followed which finally brought the railroad industry to near collapse. The dire situation finally precipitated congressional action, reversing some of the negative consequences of long-term government regulation. But even before the crash, the nation's railroads were teetering on the brink of collapse. In 1929 alone, passengers carried by Class I railroads declined over 6 percent, some 38 percent below the 1920 peak year, the number of passengers carried being about equal to that of 1905. In 1932 railroad operating revenues declined 25 percent from the year before, and 50 percent over the previous four years. And return earned on property investment was only 1.23 percent.[32] The editors of *Railway Age,* the industry's primary trade journal, summarized the situation: "Unlike industry in general, with its increasing consumer demands, mass production and a consequent opportunity to reduce unit costs, the railways are faced with a declining demand for their passenger services. The fact, accompanied by the pertinent adjuncts of retardation in the growth of freight traffic and increasing tax and wage levels, warrants . . . public policy . . . to control the development and operation of competing and subsidized forms of transportation."[33]

Besides tying the nation together as its primary transportation mode, the railroads were also a major purchaser of goods and services, and, of course, a major employer. Both were central to the nation's economic health. Between 1925 and 1929, expenditures of locomotives, cars, materials, and fuel exceeded $2 billion, but in 1932 only $612 million. The number of railroad employees declined from some 1.7 million to little over 1 million. In reporting these statistics, the editor of *Railway Age* opined that the employing and purchasing capacity of the railroads "has been reduced not only by the depression, but also, and in very large measure, by losses of traffic to carriers by highway, waterway and air that have been able to take the traffic because they are not regulated as the railways are, because they are subsidized by the federal and state governments, and because they pay their employees much lower wages for working much longer hours than the railways do."[34] Finally, the ICC allowed the railroads to accelerate abandonment of unprofitable branch lines. At the same time, new construction was essentially at a standstill. In 1932 some 1,400 miles of line were abandoned, and only 163 miles of new line constructed, the latter the least mileage reported since the Civil War.[35] Between 1933 and 1936, some 7,200 miles of right-of-way were abandoned. Between 1917 and 1936, a total of some 17,500 miles of line had been abandoned, while during the same period some 10,100 miles of new line had been added.[36]

In January 1936 *Railway Age* reported that sixteen additional companies operating a total of 29,018 miles filed for bankruptcy, joining the already swollen list

with 71,658 miles of lines operating under court jurisdiction.[37] The vast majority of the lines affected were in the Midwest and the central South. It should be
noted that the National Bankruptcy Act of 1934, in amending federal bankruptcy
legislation, had greatly sped up the process by which railroad corporations, under oversight of the ICC, could reorganize—thus partly explaining the surge in
receiverships. Also, newly created was the Reconstruction Finance Corporation,
one of the Roosevelt Administration's major depression-fighting initiatives. It
now made loans available to banks, railroads, and other institutions, enabling
them to bridge periods of insolvency. In approving rate increases through the
early 1930s, the ICC required that operating revenues first be applied to retir-
ing bond obligations. Thus revenues were diverted each year into an emergency
relief fund administered by the ICC, with monies to be paid out by the ICC to
stricken firms. The "secret" of railroad insolvency was interest on bonds, or so
wrote journalist Eliot Janeway. Interest paid on increasingly obsolete trackage,
locomotives, other rolling stock, terminals, and so forth diminished spending
on new infrastructure that was essential for railroads to truly prosper: "Thus the
very equipment which is driving the roads into bankruptcy cannot be replaced
because of the burden of debt to the bondholders. Fixed interest charges, added
to the high cost of operating the equipment on which such interest is paid, pile
up huge costs which must be met before a profit can be shown."[38] Indeed, William
Ripley (of the Ripley plan) estimated that annual bond obligations required revenue equivalent to that generated by three out of every ten freight cars moved.[39]

Congress did recognize, however, that the railroad industry was in trouble.
That recognition came as a result not only of increasing railroad bankruptcies
but also of reduced railroad purchasing and reduced railroad employment, which
contributed directly to the economic slowdown. The Emergency Railroad Transportation Act of 1933 was intended to be but a tentative first step toward averting
the industry's demise. It authorized the ICC to appoint a temporary federal coordinator of transportation who would "organize the railroads into an efficient
national system, investigate competing forms of transport, and design a program
for the extension and coordination of regulation."[40] That mandate ultimately led
to the Motor Carrier Act of 1935, which we discuss in full in our conclusion. The
1933 act brought holding companies fully under the jurisdiction of the ICC. It
brought back the idea that railroads might be consolidated, although this did
not in fact happen. It strongly recommended that interchange of traffic between
railroads be made more efficient but left how this was to be done up to the

railroads. Significantly, all requests to pool equipment, share track, and otherwise cooperate now carried antitrust immunity. The idea was to eliminate competitive duplication as much as possible. The railroads had long been charged with preserving competition to avoid monopolistic practices. Now they were not. The urgency to improve coordination was self-evident. In 1933, a loaded freight car traveling from its origin to its destination spent 84 percent of the time standing in rail yards waiting for interchange with connecting trains, and only 16 percent of the time actually moving.[41] Thus railroads were encouraged to operate joint terminals and share tracks, while streamlining accounting and reporting procedures relative to their exchange of cars.

The so-called "recapture clause" (adopted in the Transportation of Act of 1920) was repealed. It had required that half of any annual revenue exceeding 6 percent of the value of a railroad's property be turned over to the ICC, enabling the ICC to help unprofitable railroads meet bond and other financial obligations. In 1932 the Railroad Credit Corporation had been organized by the railroads themselves to pool receipts from a freight-rate increase authorized by the ICC the year before, thus to make similar loans. To pay for the new Federal Coordinator's Office, the railroads were taxed the first year at the rate of $1.50 for each mile of track they owned and $2.00 over the next two years. The office was to be closed thereafter. The act also mandated that railroad employment be held at its 1933 level and future wage rates negotiated through regional oversight boards. The act benefited the railroads most by (1) granting them permission to establish freight rate levels at "whatever the traffic would bear" (what the railroads had been variously denied since the 1880s) and (2) mandating that the coordinator investigate highways, waterways, and air service, and the need to bring them under federal regulation also. Indeed, the Federal Coordinator's Office did strongly recommend that other modes of transport, and most especially highway commerce, be strongly regulated—what railroad interests had long requested.

In the 1930s the railroads began to push back hard on Congress, hoping to liberalize federal regulations under which they operated. And they worked hard at publicizing their difficulties, hoping to engage the voting public in their predicament. However, voter antagonism toward the railroads was deeply rooted, given the industry's past abuses. Additionally, there was still fear of monopoly in the corporate economy generally and mistrust of corporate bigness for its own sake. Especially suspect were the nation's large banks and the other financial institutions that in fact controlled so many of America's railroads. Proposals to create more efficient railroading through more balanced federal oversight took decades to arrive. First, the Railroad Revitalization and Regulatory Reform Act of 1976 greatly diminished ICC authority. Then the Staggers Rail Act of 1980 largely ended federal railroad regulation altogether, precipitating railroad consolidation and merger such that the nation's Class I railroads are today consolidated into seven major systems, two of which are Canadian-based. Tens of thousands of additional miles of right-of-way have been abandoned or resuscitated as small regional and even smaller short-line railroads. Finally came the Interstate Commerce Commission Termination Act of 1996, which disbanded the ICC in favor of today's Surface Transportation Board, with its responsibility for planning and implementing a balanced transportation system based on the complementarity of various transportation modes. From the 1970s on, the railroads have found themselves better positioned to compete with automobiles, trucks, and buses. But for too long the government at various levels was oblivious to the negative impacts arising from the nation's preference for automobility. Toward the end of our period of concern, recognition had come, but it was substantially obscured by lingering government regulation and economic depression.

Chapter 2

AMERICA'S HIGHWAYS EARLY
IN THE TWENTIETH CENTURY

In 1900 the steam railroads fully dominated the intercity carrying of both passengers and freight. To the vast majority of Americans, that circumstance would likely continue indefinitely. Governmental hobbling of the railroads at state, but especially federal, levels seemed fully justified, given the railroad industry's substantial preeminence. Certainly no other means of transportation seemed about to replace freight and passenger trains. Railroading was king. The automobile, for its part, was as yet thought of mainly as a toy for the well-to-do. And yet, another industry, the auto industry, was already organizing as an effective transportation competitor. From bicycle interests had come a call for road improvement. And the electric interurban railroads were spreading their own tracks across the country. But the automobile was the true threat. Automobility in fact stood ready to revolutionize transportation in America. Highways would, indeed, very rapidly come to the fore.

How public discussion developed of the earlier claimant to public transportation, the railroad, and newer claimant, the various automotive vehicles, is a complicated matter. Railroad trade journals and some popular magazines tapped into an emotional tone underlying various data that was collected, organized, and arranged rationally. This emotional thrust was no small factor because the rivalry between automobility and railroading was very tense on occasion, short of destructive intent, to be sure—although portrayed symbolically at railroad crossings where accidents between trains and automobiles might occur—but riding on a crest of automobility's revolutionary zeal and railroading's reaction. Hence the claim to transportation's centrality in man's very existence, as an executive of General Motors Truck Company seemed to prophesy it would necessarily remain: "In studying transportation we quickly discover that it influences every phase of our existence. The transportation of goods is an absolutely essential requirement for the family, industrial and commercial life of a civilized people."[1] For the development of the highway system that was a concomitant of the twentieth century's rising automobility, the auto industry's periodicals and highway magazines joined with railway magazines to supplement the narrative about the evolving symbiotic relationship between trains and cars.

In this chapter, we explore that symbiotic relationship in two portions, namely, the rise of automobility and its sine qua non, the paved highway. The latter helped facilitate the speed and personal convenience for which automobility had begun to earn premier credentials. Automotive vehicles multiplied very rapidly. In 1900, 4,192 passenger cars but no trucks were manufactured. In 1910, 181,000 passenger cars and 6,000 trucks were manufactured in that year alone. In 1920, the numbers were 1,905,560 passenger cars and 321,789 trucks. In 1930, 2,906,082 passenger cars and 603,276 motor trucks were built; and in 1938, two years shy of the end of the early automobile age, 2,124,746 passenger cars and 530,425 motor trucks were manufactured.[2]

AUTOMOBILES

Accidents. Was it not their literal threat to harmonizing an old and a new mode of transportation, both of which were capable of considerable speed, that fundamentally challenged the promise of a satisfying and profitable cooperation? Given the relatively slow rate with which automobility crept onto the American landscape in the first decade of the twentieth century, it seems reasonable that the automobile's potential for collisions with other transportation forms was not initially preeminent. In the bustling metropolis of Chicago in 1900, a board of examiners posed only three questions about collisions when potential drivers were examined. One question would raise the issue explicitly: "What precautions would you take when approaching a crowded crossing?"[3] Competing modes of transportation, however, took comparatively quickly to blaming automobile drivers for collisions. The *Street Railway Journal* branded "the automobile [electric, steam, or gasoline] as a far greater sinner" in hit-and-run accidents "than the often unjustly condemned trolley car."[4] Plans for formally registering automobile drivers were underway simultaneously, with New Jersey in 1913 being the first state to require both a written and a behind-the-wheel examination for the applicant to earn the state's driver's license. "I confidently believe that other States will follow . . . and that the results will be fewer accidents and better road conditions," predicted the state's motor vehicle commissioner.[5]

Railroad journals countered the claims for the automobile's advantages in speed and convenience and charges of the railroad's comparative tendency toward accidents. An inspection of automobiles under the authority of the Boston police disclosed so many violations of Massachusetts law that the *Street Railway Journal* concluded that the street railway has "so often to bear the undeserved blame of accidents and a reputation for careless operation which should rightfully be laid at the door of the automobile."[6] The *Railway Age* echoed such sentiments. Not every motorist was seen as a threat, but overall, locomotives were "doing public service, urgent and beneficent." On the contrary, "the speeding motor car [was] serving the selfish pleasure" of a few while "spreading terror and suffering" throughout the growing highway system.[7]

Automobiles promised great potential to many even in the first decade of the twentieth century amid the exchanges about their accident-prone capacity. Obviously, they minimized the amount of horse manure on city streets. It was calculated that if railroads would reduce the amenities on board, they could carry more passengers, thus reducing their costs versus those of animal-drawn vehicles and thereby enhance the railroads' attractiveness. And, automobiles could serve as feeders to railroads; meshing the two technologies would thus improve service. Although meshing was a goal, automotive transportation would have to charge higher rates for freight and passengers than railroads, predicted one transportation authority on the eve of the transformation.[8] A commentator in a symposium on transportation in France took a long-range, historically based view drawn from French experience and offered that "every new invention in the way of locomotion only serves to develop a taste for travel," in the process encouraging transportation and they witnessed the automobility increasing shipments and, consequently, revenues.[9]

Observing the Vanderbilt Cup race in 1904 and 1905, which represented the most advanced automotive technology of the time, even the *Railway Age* had to

"Rigs that Run"

Single Cylinder
Balanced Motor
Water Cooled
Forced Circulation
Wheel Steering
Two Positive
  Brakes
Capacity 200 miles
Weight 1,400 lbs.
Transmission and
  Engine self con-
  tained.
Superb   Leather
  Upholstering

Patent Reachless
  Running Gear
Child's   Seat   in
  Front
Foot   Governed
  Engine
Patent Interlock-
  ing   Control
  Levers
Cylinder and Head
  Cast   in   One
  Piece
Most Reliable Car-
  riage ever built

Our Boston Model, Price, $1,200

Made especially for the eastern trade in looks, but to stand the rough wear and tear of
western roads.   Catalogue free.

ST. LOUIS MOTOR CARRIAGE CO.

1211, 1213, 1215, 1217 N. Vandeventer St.   ST. LOUIS, MO.

FIGURE 2.1. St. Louis Motor Carriage Company publicized a host of its best features in a
1903 advertisement. *Good Roads Magazine* 33 (Jan. 1903): 77.

report the remarkable speeds achieved over the one-hundred-mile distance.[10]
Such reporting continued.[11] Although the *Literary Digest* noted only five years
into the twentieth century the opinion of *Frank Leslie's Popular Monthly* that the
automobile "will become a necessity,"[12] it viewed conditions somewhat more cautiously. "Automobile enthusiasts assure us that a great revolution in transportation is upon us. Some of them even assert that it is already here. However this
may be, it seems quite certain that the motor, whether for pleasure-vehicles or
business is to play a prominent part in the world's future activities" (fig. 2.1).[13]

Those business magazines free of railroad sponsorship quickly foresaw automobiles' grand future. The *World's Work* in 1906 believed American society was
"already in sight of the time" when automobiles would "do much work of the
sort hitherto done by horses and trolley cars . . . to say nothing of the running
of two miles in a minute."[14] Later that same year, the magazine catalogued all
the automobile's uses, its author appropriately entitling the article "Automobiles
for Every Use."[15]

In a conclusion about automotive transportation not articulated during the early years when it was challenging railroads, scholars would later be convinced, as one wrote, "that [the automobile's] overwhelming commercial success was not exclusively due to the mobility it offered." It afforded both privacy and a sense of power. Railroads inherently made subjects of their passengers and even of their transporting customers because they were dependent on timetables, arrivals, and departures—conditions that the railroads set. Passengers also often sat in shared space, not in private cabins. On the contrary, "much of the appeal of the automobile stems from its ability to confer a measure of insulation from the outside world while providing at least the illusion of power."[16]

Detailed mathematical calculations ruled in public. They began to be made of the railroad-automobile rivalry as it became apparent that the new mode of transportation was becoming a serious rival. At a resort in Maine in 1908, a railroad officer counted twelve hundred automobiles conveying lodgers, many from distant cities such as Boston and New York. Automobiles, in other cases, both ate into rail travel and sometimes fed it; those paralleling rail lines siphoned off rail business, while in lateral intersection they fed rail lines. Yet, from this information the *Railroad Age Gazette* reckoned that the railroad was the loser on balance. Still, the trade journal insisted the railroad was holding its own: "And is not the enduring vitality of the railroad attested to by the fact that it survives so well not only the blows of invention, but the ruder and, in some respects, more perilous shocks of state and federal attack?"[17] Two years later, the journal calculated that both forms of transportation were close rivals where stations were near to each other, but it saw the automobile as primarily a feeder to the busiest stations, especially when the traveler carried no heavy baggage.[18] Railroads began to ask how they could take advantage of the automobile's relative advantages. They answered that on branch lines where railroads were losing money, as opposed to main lines, a motorized car on rails, or "motor car" (not to be confused with "motorcar," a synonym for "automobile"), could be added for a new or second service because the size of such vehicles could be easily reduced to carry a smaller number of passengers. The engineer W. R. McKeen, whose inventions are treated in chapter 7, foresaw that the rail motor car would "stimulate increased passenger travel and a 50 to 100 percent increase thereby is easily attained in six to twelve months, resulting in a handsome increase in passenger earnings and at the same time the freight train is scheduled to suite freight business, the cost of operation is reduced and the improved service always tends to better freight earnings."[19]

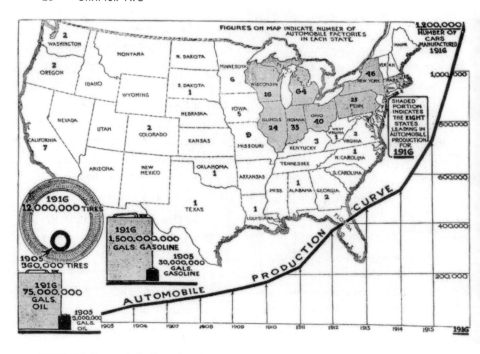

FIGURE 2.2. Concentrated heavily in the Northeast, nonetheless, many states manufactured cars. Gasoline consumption, tire production, and automobile production prove the rising popularity over 1905 to 1916. *Spokesman* 32 (Mar. 1916): 11.

Automobiles increasingly gained attention in the second decade of the twentieth century. Though not quite dumbfounded, observers of national life were nonetheless very impressed with the numbers sold and their appeal to a growing audience. In 1916 the *Spokesman* reported that 1,250,000 cars evaluated at $1 billion would be built (fig. 2.2). Of the 327 automobile factories, 99 built passenger cars. Only the high cost of maintaining a car in the city prevented even more from being bought.[20] Branch railroad lines beckoned to the growing numbers, for they could feed an area of forty to fifty miles on either side of the track.[21] Even with resources directed to wartime needs when America entered into World War I, the automobile was popular. In 1917 the Ford Motor Company, for example, made and sold 785,432 of its cars before the year ended.[22] Earlier that year it was reported that automobiles were carrying more passengers than steam railroads.[23] By the middle of the following year, motor vehicle registration had increased by little over a half million from the year before.[24]

In this second decade of the century, *Railway Age Gazette* sternly admonished bankers to pay attention to the automobile's unfair advantage since bankers were concerned with railroad finances. Automobiles had the unfair advantage of neither having to pay for road maintenance nor structures along the road. "Anyone with the price of a Henry Ford and a few gallons of gasoline in his blue jeans can enter the lists with the most costly twelve car Pullman train in the land."[25] However, the motorist should beware of repairmen who pretended to be experts in the marketplace, as opposed to "regular" blacksmiths and repairmen, because the repair business had become so profitable, warned a blacksmith's magazine, which illustrated its case in another example of the emotional appeals made in the trade literature (fig. 2.3). *Federal Traffic News* figuratively punctuated the decade in blunt words: "Verily this is the gasoline age."[26]

The problems of wartime transportation passed rapidly. War industries and the armed forces (several million men) drained off considerable railway labor, but recovery was fairly rapid. Most important, the sources of immigration that had stopped in wartime returned quickly with peace.[27] Passenger car production during wartime suffered because materials went to the military, down to 926,388 in 1918 from the 1917 high of 1,740,792; but it began rebounding in early 1919, even before the war ended, according to *Highway Transportation*.[28] By mid-1919, car and truck registrations were up by 407,791, compared to the end of 1918.[29] Automobiles drew attention not only for their functionality but also for their elegance. *Railway Age* resumed the industry's accusation of unfair competition and the charge that automobiles were "run chiefly for the comfort and pleasure of those who use them," whereas "railroad transportation is a necessity. . . . The public would be greatly inconvenienced, but would not be brought to ruin and starvation if all passenger automobiles should quit running today."[30] A *Chicago Tribune* cartoon in 1921 parodied the return of high prices, wages, and freight rates that plagued working conditions in motor transportation (fig. 2.4).

Once again, the railroad trade journals heaped blame upon automobile transportation. Even something seemingly so small as automobile taillights were a problem. While red was admittedly the color commonly used to signal danger, the *Railway Age* argued, it should not be used on automobiles. Claiming that the rear end of an automobile was not dangerous, the magazine cited the case of a driver who had followed red tail lights for a long distance and then drove around them at a railroad crossing where the gates were down and signaling red. An accident resulted when the driver drove into the train crossing on the tracks.[31] Also, in the past automotive highways paralleling railroad lines had not been problematic. Main Streets were often arranged near railroad stations for convenience but became increasingly dangerous because of increasing traffic. Railroads deserved the assistance of an alert public to battle for safer conditions, admonished *Railway Age*.[32] *Electric Railway Age* held that automobile drivers should not offer transportation to numerous friends and/or associates on the way into town and challenged the current practice of using their automobiles for transportation to and from work. Looking for allies as it slipped into defense of the swiftly outdated electric railway, the journal quoted from a New Orleans newspaper editorial: "The private automobile has its important place in the daily life of the American family, but its place is not that of a vehicle to take the owner

DON'T ALLOW THE AUTO EXPERT (?) TO SIDE-TRACK THIS TRAIN OF GOOD BUSINESS

FIGURE 2.3. "Don't Allow The Auto Expert (?) To Side-Track This Train of Good Business." The defensive blacksmith trade warned graphically of competition. *American Blacksmith* (Nov. 1912): 45.

FIGURE 2.4. Titled "On the Road to Normalcy," this cartoon illustrated the usual peacetime circumstances for the new motor industry. *Railway Age*, n.s., 71 (Oct. 29, 1921): 850.

to and from his daily work."[33] The tide of practice had already turned against the latter wish.[34] The contentious atmosphere subsided but did not end. Although absolving many individual motorists, *Railway Age* scorned the president of the American Automobile Association (AAA) in 1932 for charging that railroads favored taxes on motoring, noting that "it has been left to railroad interests to work for legislation to secure restrictions on oversized vehicles and unsafe operating practices which jeopardize the lives and detract so greatly from the amenities of private motoring." Then, the journal continued, the AAA president had rushed "into print with an entirely gratuitous criticism of men allied with the railroad industry" for working on behalf of legislation for which he, as president of an automobile association, should be expected to be the primary spokesman.[35]

With the resumption of the market untrammeled by nationalistic considerations during the war, railroads struggled with their declining revenue. The

Electric Railway Journal published the St. Louis city planning engineer's lengthy article on the necessary relief of street congestion, implicitly due to the automobile. Among fourteen solutions, the first two noted that all automobiles could not be assured street space and their parking was not a public obligation. Controlling congestion was urgent and not an aesthetic commitment. "Apparently not until the economic waste of street congestion assumes appalling proportions will the community act in self-interest," the article charged.[36] The most obvious signs on the landscape were discontinued branch lines that the rise of the automobile seemed to have rendered superfluous. By the mid-1920s, railroads were glad that the Interstate Commerce Commission permitted the abandonment of lightly traveled branch lines. In a summary of annual reports of thirteen railroads in the *Railway Age* in mid-1925, it was firmly stated that the loss of short-haul business resulted from competition with the automobile and motor bus.[37] When, only a few months afterward, the Boston & Maine was relieved of running a mere fifty-eight such lines in Massachusetts and New Hampshire, the commission said that the automobile and "other industrial forces" were the cause, and *Railway Age* charged that some of the lines should never have been built, noting that they originated amid temporary competitive advantages that were no longer in play.[38] The Bureau of Roads of the United States Department of Agriculture rendered the definitive decision in the complicated disagreement about causes, holding that automobiles were not primarily responsible. Rather, "the exhaustion of natural resources which furnished the bulk of the traffic for these abandoned lines and branches" were the cause of the decline.[39]

Discussion in the railroad world turned to broader social considerations. Just before the Bureau of the Roads declaration, the Pennsylvania Railroad's vice president stated benevolently that "the railroad and the motor vehicle [were] not rivals seeking to crowd each other out of the field of patronage" but could cooperate to find "their own broadest field in joining resources for perfecting and rounding out of the public service of transportation."[40] First, he urged, both railroads and automotive interests should abandon their contests. Second, coordination should be considered for trucks picking up freight for less-than-carload volumes in larger cities. This would raise the possibility of a trucking agency in each city. Motor bus service should be extended into outlying areas that railroads, steam and electric, did not serve. Motor trucks should handle all freight service in cities and suburbs with railroads reserved for their original purpose of carrying freight over long distances. In sum, each transportation form ought to be reserved for "the maximum service of which it is capable."[41] The president of the Illinois Central Railroad spoke for a similar strategy and tactical cooperation between railroad and automotive transportation, going so far as to encourage his audience, the American Road Builders' Association, to work for the construction of heavy-duty roads enabling trucks to laterally feed both the railroads' and the automobiles' parallel roads built earlier. Another recommended step was more construction of short, heavy-duty, hard-surfaced roads for motor truck use—but with the proviso that, most assuredly, such roads should be paid for by those who used them.[42] The market seemed to help sort through options: a report from the U.S. Bureau of Public Roads observed that passenger vehicles were used mostly on highways while railroads remained the leading freight carriers. Motor trucks predominated only in short hauls.[43] For passengers, some railroads at large stations provided the minimally priced coordinating service of auto rental; this was well received by businessmen. And for tourists who wished to avoid long drives but wanted a car at their destination, some railroads offered (for a nominal fee) to transport the car in the same train as the tourist and unload it upon arrival.[44]

The Boston & Maine, having gladly closed several of its short lines, continued operating in an area with numerous automobile highways and took advantage of that situation in the adoption of several low-rate "excursions" that earned more passengers, especially automobile drivers seeking to avoid congested roads. Defensive about a descending aura of railroading's decline, *Railway Age* denied that Boston & Maine's tactic was part of the "Railroad Renaissance" but was, rather, an element of the "greater vigor" railroads had shown since "the dark days of government operation."[45] The Pennsylvania Railroad acted upon its vice president's recommendation of 1923 and undertook a study to put into effect a productive cooperation with motor vehicles. Its president announced the results of the study in 1929. Among the findings were the following: passengers making trips partly by rail, partly by passenger bus; acquisition of financial interests in existing bus lines; and "container-car service" that enabled special containers to be used either for rail or truck transportation. The latter would save charges ordinarily levied because the railroads would not load and unload the containers. This would become a much discussed concept and increasing practice, affecting not only traffic volume but also "co-ordination" between not only automobiling and railroading but also flying.[46] At the start of the Great Depression, the strain of competition registered graphically in an advertisement by the Frisco Lines that touted travel by train because it was cheaper and more comfortable (fig. 2.5).

Exhilaration over the possibilities of automotive transportation in it several forms crested for the first time in the late 1920s. Automobile use itself developed rapidly in the 1920s. *Automotive Industries* found that 3,634,272 motorized vehicles were in use in early 1925 and that 17,726,507 were registered in the United States alone, a 17.2 percent increase in one year.[47] The *World's Work* insisted, "The Automobile has exerted a more profound influence upon our national life in the last two decades than any other factor." In hyperbolic fashion, the *World's Work* headlined an article on the eve of the Great Depression "The Conquering Automobile" and called for a better highway system in the United States so that automotive transportation could achieve its full potential; this should include "express highways leading out from the congested centers."[48] In the year after the beginning of the Great Depression, the *World's Work* itemized ways to improve automotive service, but its enthusiasm persisted: "The motorcar has firmly established itself as an absolute necessity," and car sales were taken "as good a barometer of business as we know."[49]

FIGURE 2.5. Frisco Lines both showed and reasoned why its transportation was economical, as contrasted with automobiles. *Railway Age*, n.s., 97 (Sept. 1, 1934): 260.

Statistics clearly indicated the growth of automobile use at the start of the Great Depression. The National Automobile Chamber of Commerce annually published a book of "facts and figures" about the industry. A few of the many impressive statistics showed that 20 percent of all retail sales in 1929 were automotive products, with the value of automotive products ranking it the richest industry at $3,717,996,553.[50] Even amid the Depression, motorcar registration rose in 1935 over 1934 from 2,869,963 to 4,150,000 although truck registration declined from 22,450, 000 to 21,524,068.[51] Before World War II stalled consumer purchasing, auto registration rested at almost 29,507,113, the Association of American Railroads reported.[52] Inducements to sale involved sale price, the cost of the average automobile's selling prices having declined since 1925 by more than the cost of urban living. The association also noted that automobile operating costs were cut in half of the 1926 cost by 1938.[53]

Historical studies at the end of the twentieth century dwelled on the supreme impact of automotive vehicles and granted less influence to railroads. In 1978 Robert Lieb effectively conferred victory of automotive over rail transportation. Truckers for hire steadily attracted high-value freight. Passenger automobiles provided 87 percent of intercity mileage and intercity buses exceeded passenger volume by rail. Automobiles had helped structure the rising popularity of suburban living throughout the century and the practice of individually determined social life, Lieb held. Petroleum costs, supplies, and conservation foreshadowed serious imminent drawbacks.[54] George Douglas's assessment in 1992 took a darker view of automobility, including the breakdown of long-standing and traditional social behavior and rural and small-town life. He wrote that "country stations lost their human functions, suburban stations lost their gardens to parking lots," and most important, he added, "as railroad passenger service faded, thousands of communities had no public transportation services at all." However, of the triumph of automotive transportation there was no doubt.[55] At the start of the twenty-first century, John Heitman did not even mention railroads and tied Americans' lifestyle of freedom to the private automobile: "their freedom to move is one of the most important characteristics of what they have been and who they are now."[56]

When automotive vehicles had first appeared one hundred years before, they loomed as no competition with railroads because "there really was nowhere to drive them."[57] Half the explanation for why automotive transportation surpassed the railroad requires attention to highway history.

HIGHWAYS

In the early twentieth century, when railroads ruled and automotive vehicles were comparatively new, proponents of highway building foresaw railroads to be allies. A swell of popular interest, initially in rural areas lacking adequate transportation connections to cities from which industrial commodities were distributed nationwide, largely by railroads, generated the Good Roads Movement between the 1870s and the 1920s. In *Good Roads* magazine, first published in 1892, the president of the American Motor League published a brief history of highway improvement since the late nineteenth century as prelude to an extension.[58] The benefits would be reciprocal to roads and railroads, it was imagined. *Good Roads Magazine* in 1901 explained that an improved common highway "means an increase in production, and consequently, more business for railroads. It means that any product may be hauled to a railroad at any season of the year, regardless of weather conditions."[59] Only a month earlier, the *Railway Age* cited the case of the Iowa railroads, which asked the state for low rates on crushed stone and gravel to build and improve country roads, and referred to "many railroads" that agreed, noting that some "good roads trains" were sent out to persuade the public of the value in road building.[60] At the onset of railroad support for automotive highways, the *Railway Age* eagerly reported early steps taken for the Brownlow bill before Congress to finance model highways in states and territories to promote the future cause of highway building. Under its terms, $20 million was to be assigned to highway construction.[61] In 1916 *Good Roads Magazine* cited an example of the growing railroad support for road building—that of the Nashville, Chattanooga & St. Louis Railway for the Good Roads Movement in Tennessee.[62] Immediately upon America's entry into World War I, *Good Roads* especially promoted road improvement and cited the railroads' long-standing support: "They are essential—essential to commercial development, essential to military successes, and essential to social, intellectual and moral progress."[63] England's longer wartime in World War I stoked *Highway Transportation*'s zeal for road building on military grounds, implicitly making construction a matter of life and death.[64]

Highway construction plans throughout the states resumed quickly after the war's end.[65] One of the most important legal steps influencing those plans was a decision in 1921 by the U.S. Supreme Court in favor of a petition by the Kansas City Southern and the Arkansas & Ft. Smith Railroads against an attempt in an Arkansas road-improvement district to assess those railroads for construction of roads the district assumed would increase the railroads' value. *Railway Age* argued after the court's decision that levying a tax in excess of that on the current value of owners' property neighboring the roadway was wrong because it was doubtful that the road would enhance the railroad. The trade journal sternly insisted that "the railroad companies have not been treated like individual owners, and we think the discrimination so palpable and arbitrary as to amount to a denial of the equal protection of the law."[66] In Illinois, the Chicago, Rock Island & Pacific filed with the state's utilities commission a similar protest against motor buses using hard roads parallel to the rail lines as private gain from public improvements that were profitable because of the rail line.[67] Reverting to the former assumption of the railroads' superior management, as compared with that of public road managers, *Railway Age* held that the latter should exercise similar competence. "This will involve far more engineering study and design than has been given most highway construction to date," the trade journal demanded.[68]

Road-building interests believed that events in the final years of the second decade of the twentieth century would justify the federal government's support of automotive road building as opposed to railroad building. One of the most important points, reasoned the author of a letter to the editor of *Good Roads Magazine,* was that the principal construction materials for highways were less expensive than that of railroads—limestone, clay, shale, and cement versus iron and steel.[69] Another contributor to *Highway Transportation* advised that roads should not be built cheaply but to bear, around the year, the weight of the load passing overhead: "we must build the road to carry the load." That was identical to the railroads' circumstances fifty years before.[70] *Highway Transportation* judged that railways were unfit to be the sole transporter; rather, they served "as connecting links between centralized portions of the population." It was aggressively concluded that "there is altogether too much blind confidence in the power of the railways as the sole transportation medium of the country." A single transportation means, railways were "not the framework and body" of the system.[71]

Railroad men and users of highways for commercial gain continued arguing what should be built and who should pay. The public discussion became immersed in complicated exchanges. *Highway Transportation* launched a series of articles in the early 1920s stating the exact specifications for various highway types. Trunk highway surfaces, for example, should be built with a minimum of twenty-foot-wide pavement. The articles read very much like technical engineering reports.[72] Railways had promoted highway construction earlier but, since the enactment of the Federal Highway Act of 1921, it seemed that "irresponsible men and concerns" who "bought trucks on the installment plan, that are unregulated as to rates or character of service" paid little toward highway maintenance, *Railway Age* editorialized. Yet, in a conciliatory frame of mind, the argument cropped up again for coordination and consideration of the greater good, not the supremacy of a particular transportation means: "The hard road is here to stay. It is an agency of transportation with a definite place. The problem is to so co-ordinate it with other means of transportation that it may fill this place most satisfactorily and contribute most directly to the welfare of the country."[73] Not all Good Roads advocates thought railroads performed the lesser of the two services. In 1922 the president of the state of Washington's Good Roads association declared the following: "Electric and steam railroads are the only safe, sane and permanent means of commercial traffic."[74] This sounded much like an editorial in the *Illinois Central Railroad* magazine, which stated that "a great public service" was rendered by railroads, whereas those using highways for freight and passenger transportation were in business for private profit.[75] A more balanced public policy, rather than a preference for one mode of transportation over another, seemed to be making progress, however slowly.

A continual debate, however, raged through the twentieth century. Were freight and passenger automotive services paying their fair share for the highways on which their business depended? How much should railroads pay for highways, if anything? Even to a degree, although lessening over time, were railroads essential to America's economy and even culture? Transportation analysts and historians alike concur on the following points. Except for development of the interstate highway system from mid-century, the nation's highway system was completed by the late 1920s.[76] In the century's first forty years, the volume of public streets and highways grew incredibly, from 8,000 miles in 1900 to 29,507,113 in 1941. Railroads, steam and interurban electric, lost much of their passenger service to the automobile and bus. Trucks supplanted horse-drawn vehicles. Commercial automotive vehicles complemented rail and other carrier means.[77] The Interstate Highway System, begun in 1956, constituted the most significant highway system following World War II, with intercity buses carrying more passengers than other systems. By the late 1970s, trucks carried 21 percent of intercity-freight ton mileage.[78] Railroads were reckoned to be three to four times more fuel efficient.[79] They still had a place and especially so in an increasingly environmentally conscious world.

While railroads were stabilized in public opinion, automotive transportation faced its own decline, as had railroads earlier. James Flink, the title of whose history of the automobile underscores its importance in American life, *The Automobile Age*, judiciously traced the brief period in the early 1970s when "death of the automobile" literature sprang up. Oil shortages, the automotive production industry's foolish styling and questionable marketing policies, and the competitive rise of Japanese products gave incentives for more public transportation and fuel economies.[80] Steps for recovery positioned automotive transportation in the lead again, Flink believed, if only for a while; the half-century of the nation's development deserving the title of the automobile age ended in the 1970s.[81] His key periodization of the automobile industry in the 1970s is now universally accepted.[82] Nevertheless, cars, among all the varieties of automotive vehicles, topped American culture. Together, literary scholar Ronald Primeau wrote, "Roads and cars have long gone beyond simple transportation to become places of exhilarating motion, speed, and solitude."[83]

THE IMPORTANCE OF AUTOMOBILE-RELATED INDUSTRY IN THE NATION

From the early 1920s through the early 1950s, the United States dominated the world's economy, and its largest industry was the automotive industry, principally because of the mass-production process the industry had perfected. Henry Ford was the foremost creator of this system.[84] *Automobile Topics* in 1918 reported that car manufacturing, parts and accessories businesses, and dealers and garages combined to make the nation's third largest industry. Some 280,000 automobile workers were employed. Parts and accessory businesses employed 253,000 workers. Registered motor vehicles totaled 4,842,139, of which 435,000 were trucks.[85] These numbers were comparable in 1927.[86] The total number of registered motor vehicles had jumped to 24,493,124.[87] The National Automobile Chamber of Commerce in 1931 reported data from the year before that justified the claim for "a continued growth in motor transportation although at a slackened pace" as the Great Depression set in.[88] On the eve of the Great Depression, the same source recorded that since 1916 records of the U.S. Bureau of Public Roads showed that highway bonds had yielded $740,614,500, while another $372,243,091 for highways had come from federal funds.[89] Two years into the depression, 3,030,000 miles of roads existed, and 128,000 miles were "high type surface."[90]

In addition to this huge public expense, private profit had given reason for the founding of numerous businesses supporting automotive transportation. In 1929 these wholly new forms of free enterprise included 3,379 body, fender, and paint shops; 60,627 repair and storage, gasoline, oil, and accessory shops; 2, 059 stations, garages, and lots for parking alone; and 728 radiator shops, including those for repairs.[91] In pointing out that increased automotive vehicle registration produced increased demand for maintenance and service, *Motor Age* tabulated in 1940 that car and truck registration had increased four times over the previous year.[92] Post–World War II prosperity by 1960 resulted in the wholesale sales value of motor vehicles at $14,461,150,000.[93] In 1950, while the United States' population was only 5 percent of the world's, it was producing three-fourths of the vehicles in the world and owned two-thirds of the world's vehicles.[94] American domination of the automotive industry, however, crested in 1955, when the import market began.[95] In the final decade of the twentieth century a global market developed, as foreign cars sold in the United States and U.S. brands with parts manufactured abroad were sold abroad.[96] The early twenty-first century witnessed a serious decline; employment in the U.S. automotive industry maintained conditions only slightly better than other domestic industries.[97]

The heart of the exchange between the nearly century-old railroad system and the rise of automotive vehicle transportation depended on how those involved in two systems debated each other and how the systems performed in public. Federal and state legal systems and regulatory agencies were important, but they provided a context in which the public debate unfolded and at times raged. After the first twenty years of this exchange, it became clear to most participants and public bystanders that the railroad and automotive systems each had their own role to play in American life and the nation's economy. A world war and a depression in the 1930s fueled the ongoing debate. On the eve of World War II, however, this debate illuminated nothing new. The American automotive industry outshone all other American industries until mid-century. Both railroads and automotive vehicles clearly were here to stay. Still, the unfolding rivalry spawned a series of important issues.

Chapter 3

THE RAILROADS ADOPT
AND ADAPT

The railroads sought in various ways to adjust and adapt to the rapidly increasing popularity of motor vehicles. They were especially quick to employ internal combustion engines and electric motors in ways similar to what the automakers had achieved. Track inspection and maintenance, for example, was "motorized" with self-propelled vehicles of different descriptions replacing light hand cars, but heavy equipment, such as cranes previously pulled by locomotives, was also introduced. Automobiles represented an important new kind of freight. Trains carried both fully assembled automobiles and auto parts, the former handled in specially designed freight cars. Petroleum was carried in bulk, including crude oil and finished products such as gasoline, and the railroad tank car quickly became a common sight. Picturing automobiles figured increasingly in railroad advertising, with the ability to rent cars at one's destination, for example, promoted as a means of increasing travel convenience by train. The increasingly ubiquitous taxicab and the increased use of private cars dictated that passenger depots be made fully automobile-convenient with drive-up entrances and, of course, with nearby parking facilities. Around 1900 several railroads sponsored "Good Roads trains" that carried road-building equipment and their operators from locality to locality for demonstrations that promoted road improvement. Improved rural roads would greatly facilitate the movement of farm produce and other commodities to railroad sidings, it was argued. "Rail highways" were briefly considered with automobiles, trucks, and buses moving along rails. But it was something that did not catch on. Henry Ford, the nation's most successful automaker, aggressively suggested how the railroads might be better built and better managed, buying a railroad to demonstrate his thinking. The railroads steadily increased their opposition to government regulation and taxation. Intensive lobbying in state legislatures and in Congress emerged. If the railroads had to be regulated and taxed, it was increasingly argued, then other transportation modes needed to be regulated and taxed equally. Such are our concerns in this chapter.

"MOTORIZED" TRACK INSPECTION AND REPAIR

Inspection cars, usually antiquated passenger coaches pulled by locomotives, were used to inspect tracks on steam railroads. On interurban lines it was usually a retired passenger car. For decades section hands had gone out on hand cars to tend to regular maintenance and make repairs. Over time they were replaced by motorized versions. Another first step, however, was to adapt ordinary motor vehicles to run on rails. "A New Inspection Carriage" was the headline on a 1902 article in *Railway Age,* which read in part: "Stripped of its pneumatic tired wheels and provided with another set having inside flanges, the automobile—the name is used for lack of a better—may become a light inspection car, businesslike in its behavior and luxurious in its appointments."[1] One typical inspection car was built in 1904 by Oldsmobile, the company having just been bought by General Motors and moved to Lansing, Michigan, from Detroit. "The car is built for standard gage, has a 62-in. wheel base, oak sills, 20-in. pressed steel wheels . . . cold-rolled steel axles, Hyatt roller bearings and powerful brakes of the expanding clutch type. These cars have ample capacity for water and gasoline to run the car 100 miles or more, and the cost of operating is very slight, as a gallon of gasoline is sufficient to drive it from 20 to 25 miles," *Street Railway Journal* reported. It carried upwards to eight passengers. It was essentially the Oldsmobile Runabout put on rails through special wheel attachments.[2] The car began General Motors' involvement with the railroad industry, the company becoming a leading manufacturer of diesel locomotives in the 1930s. The four-passenger Stover Motor Car, built at Freeport, Illinois, was specifically designed (or, perhaps more appropriately, redesigned) for "the use of general managers, superintendents, engineers of maintenance of way, roadmasters, water service men, bridge and signal engineers, and others having inspection of track as part of their duties."[3] A larger twelve-passenger vehicle followed the next year (fig. 3.1).

By 1913 at least thirty-one Class I railroads were using specially designed "rail motor cars" for maintenance work, manufacturers having added tool boxes as well as open cargo-carrying spaces. Cars were outfitted with spike-driving and other power tools that greatly reduced labor while at the same timespeeding up work. "The argument in favor of motor cars in section service," *Railway Age* reported, "are that the men arrive at the work without the fatigue due to pumping a hand car and can spend the whole day working on the track. Gangs can be combined more quickly in case of emergencies. Men can be more easily secured and held in section work when more cars are provided. On account of the saving in time and energy, the length of sections can be increased."[4] The Signal Department of the Norfolk & Western Railway had some seventy cars in operation as early as 1911. According to the *Railway Age Gazette,* "Officers of this road estimate that

FIGURE 3.1. Track inspection car. "Stover 12-Passenger Motor Car," *Railway Age,* n.s., 45 (Feb. 7, 1908): 204.

A Motor Car is a Stepping-Stone To Advancement

From Roadmaster to Superintendent

FIGURE 3.2. Advertisement for Burton W. Mudge & Co., Chicago. *Railway Age Gazette,* n.s., 57 (June 21, 1912): 19.

the resulting savings in labor has been 25 per cent, while the increase in efficiency has been 50 per cent."[5] Of course, it was the hand car that was traditionally used for track-repair work. They had the advantage of being much smaller and thus much lighter than motor cars. They could easily be lifted from the rails on the approach of a train. Although they could carry neither as many workers nor as many tools, light trailers could be attached to enhance carrying capacity. Hand cars were motorized, ceasing to be hand-propelled in the process. The Burton M. Mudge Company of Chicago produced one version (fig. 3.2).

MOTORIZED BAGGAGE AND FREIGHT HANDLING

Self-propelled or pulled by tractors, baggage carts at passenger terminals and freight wagons at freight stations were also "motorized." With electric motors battery-powered, rather than with internal combustion engines, they greatly facilitated the loading and unloading of trains. Self-propelled baggage carts could be operated from either end since devices for steering and acceleration were duplicated, two narrow platforms being provided on which drivers could stand. Turning carts on narrow freight or passenger platforms was thus not necessary. Tractors, on the other, moved one way and were provided with seats for drivers. They required wide spaces in which to operate. Batteries for both, which required frequent recharging, could be easily slipped in and out with little disruption of service. Pictured is a baggage-cart tractor that was manufactured, like the Oldsmobile, at Lansing, Michigan (fig. 3.3). The illustration, from a *Railway Age Gazette* advertisement, was accompanied by the following text: "The 'Lansing Way' relieves congestion and saves from $50.00 to $100.00 per day in trucker's wages. The 'Lansing Train' rolls up to the freight car door, receives loads for each trailer, and the freight goes on its way at automobile speed."[6] (It should be noted that every piece of motorized equipment used for baggage handling tended to be called a "truck" and a string of carts a "train.")

Motorized off-rail vehicles gradually came to affect railroading. The editors of *Railway Age* offered this opinion in 1931:

> In the industrial tractor, lift truck, crane truck and highway truck, and other trackless power units for handling materials, the railways have useful and effective equipment with which to reduce costs. Material-handling costs, freight-handling costs, rail-transportation costs and the cost of doing maintenance and construction work are all vulnerable to this equipment. . . . Because of its strength, speed and flexibility, such equipment is revolutionizing railway storehouse, repair-shop and also freight-house practice. It puts manual methods of handling material on the defensive and forces the warehouse truck and wheelbarrow, as well as the rail push car, into the past as surely as the automobile has pushed the horse off the highway.

Such equipment was, of course, much improved by the 1930s, and there was great diversity:

> There are tractors equipped only to pull; others have boxes for loads; still others are equipped with large wheels for high speed, or with crawler treads for rough ground; some are equipped with bumpers for pushing; while scores of different accessories are manufactured for special uses, including especially the various crane attachments for hoisting and lifting, and power brooms and plows for cleaning roads or moving earth. The trailers also vary widely in types and sizes. Dump wagons, demountable bodies, oil tanks and other special equipment are included.[7]

In 1932, a survey of American railroads established the following:

FIGURE 3.3. Battery-powered tractor. Advertisement for the Lansing Co., *Railway Age Gazette*, n.s., 61 (June 30, 1918): 33b.

> The company-owned equipment engaged in company work . . . includes 61 motor buses, 374 passenger automobiles and 1,070 auto trucks, together with 302 auto-truck trailers. This does not include numerous ambulances, fire engines and other specially-constructed machines. The equipment also includes 1,114 non-burden-bearing and 2,164 burden-bearing industrial tractors, 133 tractors equipped with hoists and 12 tractors with crawler treads. . . . The railroads also own 560 electric crane trucks, 4,234 self-elevating trucks and 567 motorized warehouse and baggage trucks. . . . Additional equipment includes 430 jack lift trucks . . . together with 36,995 trailers for industrial tractors.[8]

Automobiles, it was reported, were utilized by freight solicitors, lumber inspectors, and others to reach points not served by a railroad as well as to "expedite the movement and otherwise facilitate the work of engineering parties, line men and supervisors' forces."[9] The Pennsylvania Railroad was the largest operator of industrial tractors with 159 units.[10] These tractors, powered by internal combustion engines, were, in a sense, the precursors of the diesel switch engines that, beginning in the late 1920s, came to dominate railroad-yard work in the 1930s. It was in railroad yards that "steam railroads" first stopped using steam engines and thus became, in a sense, fully "motorized."

Especially where railroad lines were paralleled closely by highways, or could otherwise be accessed readily by road, trucks substituted for work trains, bringing work crews and their tools to bear on track and other repair problems with speed and efficiency. Trucks delivered coal to heat company facilities, and hauled away the ash. They delivered supplies of "less-than-carload" implication: package freight of all kinds that moved increasingly by truck from railroad box cars to customer destinations. Electric railroads with their overhead wires had a special need. Line trucks (sometimes called tower trucks) featured scaffolding with which power lines could be strung and repaired (fig. 3.4). Trucks manufactured by the White Company of Cleveland, Ohio, were specially outfitted for use by streetcar and electric interurban railroad companies; they featured power

FIGURE 3.4. Union Street railway line truck at New Bedford, Massachusetts. *Autocar Trucks As Servants* (Ardmore, PA: Autocar Co., 1927): 2.

winches used in pulling both aerial and underground cabling, hauling and setting poles, hoisting transformers, pumping out manholes, righting derailed cars,
and clearing track of obstructions. "Supply bins for carrying small material are
mounted on the side, as are also special brackets for carrying ladders, pike poles,
etc. In addition, ample room is available in the body of the wagon for carrying
wire and other bulky material," the *Electric Railway Journal* reported.[11] Thus motor trucks helped railroads cut maintenance costs in various ways.

AUTOMOBILES CARRIED AS FREIGHT

In November 1900 a short news items appeared in the *Railway Age* headlined:
"Automobiles are not baggage." The Central Passenger Association had ruled that
travelers with motorcars would not be allowed to check them for free as a kind
of luggage. True, most cars in 1900, although motorized, were still little more
than quadricycles—lightly built vehicles with four rather than two wheels. "The
capacity of baggage cars is limited, and so a line of distinction has to be drawn,
with bicycles and baby carriages admitted and automobiles and the rest ruled into
the freight or express car."[12] Over the next decade the railroads adopted specially
equipped freight cars for automobile transport, not to accommodate motorists
traveling by train but to accommodate automobile manufacturers in distributing their product. Most cars included a door at one end for end-loading, and
extra-wide side doors for side-loading. Many featured double deck frames, two
automobiles being hoisted up high with two below (fig. 3.5). Motor cars were an-
chored in place with heavy straps clamped in frames and/or nailed into flooring.
Wood blocks were used to further stabilize vehicles. Special loading platforms,
ramps, and even cranes were used to lift autos, the latter employed when box-
car shortages necessitated use of gondolas and flatcars. In 1920 the Buick Motor Company of Flint, Michigan, used railroad freight cars to ship out some 40
percent of its production.[13] Railroads serving Detroit and southern Michigan
(particularly the Pere Marquette, Grand Truck Western, and the Michigan Central) benefited greatly from auto manufacture in and around Detroit. In 1924 the
latter railroad handled over 163,000 carloads of automobiles and trucks, which
constituted some 15 percent of its traffic measured in carloads and some 4 percent
of its total tonnage.[14] In 1927 American railroads loaded some 757,000 carloads
of newly assembled automobiles. However, in total some 3.2 million carloads
could be traced to both the manufacture and use of automobiles, including car-

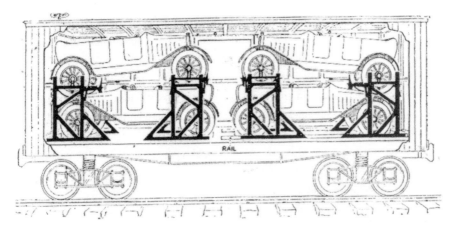

FIGURE 3.5. Freight car configured as automobile carrier. "'Carbo' Steel Decks for Carrying Automobiles," *Railway Age*, n.s., 69 (July 16, 1920): 108.

rying of petroleum products. In contrast, the railroads loaded 17.4 carloads of mine and forest products (with coal leading the way) and 13.3 million carloads of merchandise or "less-than-carload" freight.[15] But with the arrival of the Great Depression, the number of carloads traced to auto manufacture dropped to 1.9 million carloads.[16] Nonetheless, it remained important. In 1938 trucking accounted for some 55 percent of new car haulage, the railroads some 38 percent, and waterways (especially on the Great Lakes) some 7 percent.[17]

Several railroads began to rethink the carrying of cars not as freight but as "baggage." The Southern Pacific put in operation in the late 1920s several specially designed baggage cars to haul the cars of passengers headed from San Francisco to such destinations as Yosemite National Park and over the Sierra Nevada Range to Reno. The intention was to "attract considerable additional business which would otherwise be lost to the highways."[18] Eventually called "train-auto" service, the cost for the service was five first-class, one-way adult passenger tickets (three for the automobile and two for the driver and one passenger).

PETROLEUM

With the internal combustion engine winning out over steam and electrical motors (both steam and electric cars had limited driving ranges), carrying crude oil and refined products such as gasoline and lubricating oil became an important source of railroad freight revenue. Thus tank cars—indeed whole trains of tank cars—became commonplace. In 1927 American railroads carried over 1 million carloads of gasoline, 80,000 carloads of crude petroleum, and 45,000 carloads of lubricating oil. In addition, they carried some 44,000 carloads of asphalt for road construction. The latter compared with 171,000 carloads of cement and 782,000 carloads of sand and gravel, both used largely in road construction. Of all manufactured products carried by rail, refined petroleum ranked first; bar and sheet iron, structural iron, and iron pipe (much of the latter used in petroleum pipelines) ranked second; and automobiles, trucks, and motor-vehicle parts ranked third.[19] When new oil fields were opened in the 1920s, such as the Seminole Field in Oklahoma and the Bolger Field in Texas, railroads handled most of the output, at least until pipelines were completed. In one year the production of the Seminole field reached some 300,000 barrels of crude oil a day, requiring the Chicago, Rock Island & Pacific Railroad to improve its Choctaw Line (which ran from Memphis to Tucumcari, New Mexico) with scores of miles of passing track and a massive new yard at Seminole. In one five-month period in 1927, some 12,800 loaded tank cars departed outbound from Seminole. Some 7,000 tank cars were handled inbound. In addition, the railroad had brought to Seminole most of the equipment and matériel necessary to create the oil field and to enlarge Seminole as a town. And when construction on a pipeline did begin, it brought the pipe. On average seventeen freight trains operated daily in each direction.[20]

At Borger, Texas, the quintessential oil boomtown of the American twentieth century, similar demands were placed on the Atchison, Topeka & Santa Fe Railway. In less than a year, Borger grew from nothing into a town of some thirty thousand people. The Santa Fe built a branch line from Panhandle about thirty miles away. *Railway Age* reported: "The station at Panhandle, Texas . . . whose freight traffic has hitherto consisted almost exclusively of wheat and cattle shipments, has within the past year become almost equal to Chicago on the Santa Fe system in earnings." The movement of tank cars, loaded and empty, into and out of the oil field, amounted to some 134,000 cars in six months. Some eleven hundred rigs were pumping crude oil, each rig having required approximately five carloads of pipe, lumber, steel, and machinery in its construction.[21]

Until 1929 pipelines in the nation were almost completely given to transporting unrefined petroleum and natural gas. Several gasoline pipelines were then put in operation, the result being that eventually the railroads lost much of their gasoline traffic. Some oil companies had been quite dependent on the railroads to move crude oil from producing fields to refineries, and from refineries to market. For example, approximately 97 percent of the Pure Oil Company's outbound traffic from West Virginia had moved by rail in leased cars, the number of cars used increasing from seventeen hundred in 1921 to twenty-six hundred in 1928. According to *Railway Age,* gasoline production in the United States between 1917 and 1929 increased 546 percent. Overall railroad tonnage kept pace, rising from some 21 million tons in 1917 to some 46 million tons in 1929. Refined petroleum, for its part, represented about 4 percent of the carload traffic nationwide, and produced about 6 percent of the total operating revenues.[22] But the nation's eighty-six thousand miles of pipeline in use then were not only more efficient at transporting petroleum products but also substantially cheaper at doing so. Nonetheless, in 1935 America's railroads were still earning some $200 million dollars annually from transporting petroleum products.[23]

THE GOOD ROADS TRAINS

In November 1900 road enthusiasts from thirty-two states met in Chicago to establish the National Good Roads Association. The "Good Roads Movement" was initiated by the League of American Wheelmen. With a membership in the thousands, concentrated mainly in the American Northeast, the organization lobbied for hand-surfaced roads in and around major cities. Also represented at the meeting were officials from newly organized automobile clubs, government officials from the Department of Agriculture and U.S. Postal Service, and, as well, representatives of the nation's railroads. Better "wagon roads," railroad officials believed, could only benefit their operations. Not only might freight volumes climb, but operations might also be improved. Muddy roads in wet periods slowed the movement of farm produce to railroad sidings, thus creating uneven demand for rail cars and making train scheduling problematical. Better roads would also increase market radius and thereby enable a given railroad to encroach on territories of competing lines. Better roads, as a stimulus to better farming, would generate new business over the long term. For the short term, railroads stood to benefit from the shipment of road-building materials and equipment. Next to organizing the association's work, which included plans for a monthly magazine, perhaps the convention's most important accomplishment was to propose a "Good Roads Train." It was to be run "on one of the great railway systems out of Chicago, which should traverse its entire line, stopping at central points for the holding of district and state conventions and the construction of object-lesson roads."[24]

The Illinois Central then volunteered at no cost to operate the first train the following year, north from New Orleans to Chicago. And then the Lake Shore & Michigan Southern Railroad offered to run a second train, and the Southern Railway a third. The Illinois Central venture operated over three summer months serving sixteen road-improvement conventions held in major towns and cities along its main line. The train carried road-building equipment donated by various equipment manufacturers, with men supplied to operate them. For example, a rock crusher, a roller, and diverse plows and scrapers were carried. A Pullman car was provided for railroad and other officials, and dormitory cars for workers. In addition, several gondolas carried stone, brick, and sewer pipe. Sections of road up to a half-mile in length were variously improved at each stop. Thereby, local convention participants were shown, using gravel and other materials locally available, not only what was possible by way of road improvement but also how it might be accomplished. Locals were instructed "by practical example" how roads were "tiled, drained, topped and finished."[25] The Lake Shore & Michigan Southern rebuilt several miles of road near Buffalo, New York.

The Southern Railway's excursion, however, was probably the most ambitious, being combined with eighteen state and district conventions over a five-month period. Again, at each stop sample roads were built. The train carried "a number of road graders, horse rollers, 10-ton steam roller, and two elevating graders . . . two traction engines and three crushing plants . . . [and] distributing cars to surface the earth road with stone, gravel, chert, slag, tar, macadam or oyster shells, such as the various communities furnish."[26] The Great Northern Railroad in 1902 ran a train across Minnesota, North Dakota, and eastern Montana (fig. 3.6). Although these Good Roads trains operated only for a short period, railroad participation very much jumpstarted the Good Roads Movement. The organization went on to hold annual road conventions at state and local levels; lobby for government road-building programs at county, state, and federal levels; and otherwise popularize the idea that motoring in America was not merely a passing fad for the well-to-do, but an important new form of transportation for Americans generally.

The railroads helped in other ways. Many donated road-building materials, hauled those materials at reduced freight rates (and in a few instances for free), and even donated abandoned railroad rights-of-way for highway development. Pictured is the new Nicholson Bridge on the Delaware, Lackawanna & Western near Scranton, Pennsylvania. The railroad's former right-of-way, shown below, was donated to the state of Pennsylvania and reconfigured as a highway dubbed the "Lackawanna Trail" (fig. 3.7). In straightening its main line across Wyoming, the Union Pacific also donated abandoned right-of-way for improving the closely paralleling Lincoln Highway. Numerous ideas were advanced whereby the railroads might have played an even greater role in highway development. Out of the U.S. Department of Agriculture in 1904 came the suggestion that rural roads be provided with steel tracks, thus to carry both horse-drawn carriages and wagons and motor vehicles. Under this plan, a road's entire surface would not need improvement, thus reducing costs. Indeed, a proposal to create "automobile railways" across Iowa was taken to, but then defeated by, the state legislature. Rails would have been from eight inches to a foot wide, with flanges two to eight inches high in order "to keep the wheels of the motor cars from running off."[27] In 1932, there came from the Rhode Island Commission of Foreign and Domestic Commerce a far more practical suggestion. The commission

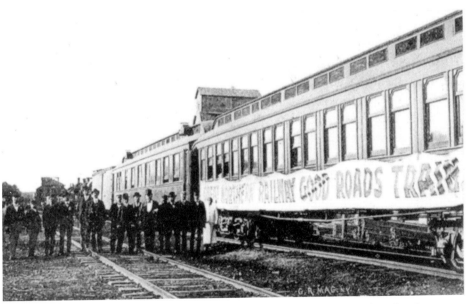

FIGURE 3.6. The Great Northern Railroad's Good Roads train. "The Great Northern Good Roads Train at St Cloud, Minn.," *Good Roads Magazine*, n.s., 3 (Oct. 1902): 1.

FIGURE 3.7. Nicholson Bridge and Lackawanna Trail near Scranton, Pennsylvania, circa 1920. Postcard, authors' collection.

estimated that some two thousand of the eight thousand miles of railroad track in New England had been rendered obsolete by changing trends in transportation, including the introduction of motor trucks and buses. "The commission believes," reported *Bus Transportation*, "that if obsolete track was turned into productive heavy-duty highway at least part of this investment could be salvaged. The railroads might keep such highways exclusively for their own trucks, but could charge tolls for motorists. Or they could, perhaps, sell their rights-of-way to the respective states.[28]

TAXICABS AND RENTAL CARS

Although they did little to promote the use of taxicabs in the United States, the railroads nonetheless depended heavily upon the cab industry. Travel by train would have been exceedingly difficult without cab service at points of passenger departure and arrival. Taxi service was an enterprise that had matured during the horse and carriage era. In the auto age, the horse-drawn cab logically gave way to its motorized cousin, the taxicab. However, the costs of buying, servicing, and operating motorized vehicle fleets significantly changed industry ownership patterns. Largely gone, at least in America's cities, were the "mom and pop" operators. Brought to the fore were large corporations, many attaining national reach by the end of the 1920s. Municipal licenses were fought over, and in some cities, such as Chicago, heated battles were physically fought. In 1931 there were some 81,000 taxicabs in the United States out of some 26.5 million automobiles registered. They accounted for about 1 percent of all motor vehicle traffic. But in the nation's largest cities, statistics were quite impressive. Taxicabs in New York City accounted for less than 3 percent of the city's registered vehicles, but some 32 percent of the vehicle mileage.[29]

Perhaps, the most successful American taxicab entrepreneur was John Hertz, who in 1907 began in Chicago with several used cars. He founded in 1915 the Yellow Cab Company, yellow being the signature color of his vehicles. In 1920 he established the Yellow Cab Manufacturing Company to assemble his own cabs, and in 1924 the Yellow Coach Manufacturing Company to manufacture buses.[30] Yellow Bus was later bought by General Motors. Chicago was the nation's major railroad center. Unfortunately, most travelers passing through the city were required to change railroads, which often meant changing passenger stations as well. Not only did taxicabs enable such transfers but limousine service did so as well—particularly that of the Parmelee Company, which John Hertz bought. The company was founded in 1854 as a baggage-transfer operation. In 1908 it opened a cab service but restricted it largely to serving the city's railroad depots, with motor vehicles rather than horse-drawn carriages being used for the first time.[31]

Not only did taxis come and go to the nation's railroad stations, but travelers did as well in their own autos. Around large-city railroad stations automobile traffic quickly became a nuisance, given the lack of curb space both for dropping off and picking up, and, more important, for parking. In small towns the auto, although less of a problem, was still something to be managed. Pictured from about 1920 is the "Big Four" (Cleveland, Cincinnati, Chicago & St. Louis Railroad) depot at Mattoon, Illinois. Taxis and cars are edged right up to the station's passenger platform (fig. 3.9). When new terminals were built, they necessarily included drive-up entrances and, of course, off-street parking. Big cities sometimes had streetcar platforms integrated into railroad stations. When the Pennsylvania Railroad opened its new downtown Newark, New Jersey, terminal in 1935, a separate bus facility was located one level up from the transit platforms, providing for "close co-ordination with trolley, bus, taxicab and private vehicular traffic."[32]

The railroads might have flirted off and on with passengers shipping their cars as "baggage," but they definitely supported rental cars being available. Among the earliest rental car firms in the United States was the Saunders System, begun in Omaha, Nebraska, in 1916. Originally, it was named the Ford Livery Company for the brand of automobiles it used. Eventually the firm grew to some ninety rental stations in sixty cities in twenty states.[33] The Hertz "Drive-Ur-self" system originated in Chicago as a logical outgrowth of John Hertz's Yellow Cab Company. A coast-to-coast chain of franchised rental stations was quickly created largely through franchising, with licens-

ees operating what at first were referred to as "gasoline livery stables." Originally, neither Saunders nor Hertz oriented their locations to railroad depots but rented their cars from central business-district locations. Nonetheless, cars could be delivered to railroad depots, and thus readily picked up at train depots. Their prime customers initially were salespersons and others traveling on business.

In addition to gasoline and lubricating oil, and expenses due to car depreciation, tire replacement, repair, insurance, and so forth, the per-mile cost of running an average-priced car was at minimum about 10 cents a mile in 1927. The average cost of railroad travel was some 2.8 cents per mile. For salesmen, for example, it made sense to ride the rails between principal points and then rent cars to visit customers.[34] In 1940 several railroads (including the Burlington, the Milwaukee, the Chicago & North Western, the Santa Fe, and the Union Pacific) established the Train Auto Service, Inc., in partnership with Railway Extension, Inc., a company created to supply rental cars from auto dealers in cities

FIGURE 3.8. Advertisement promoting train service to Yellowstone National Park, Burlington Railroad, Circa 1925. Advertising brochure, authors' collection.

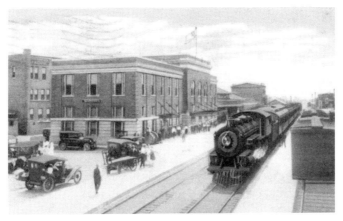

FIGURE 3.9. Big Four Railroad Station, Mattoon, Illinois, circa 1920. Postcard, authors' collection.

FIGURE 3.10. Advertisement for the "Drivurself" Rail-Auto Travel-Plan, 1940. Advertising brochure, authors' collection.

across the nation (fig. 3.10). "A railroad passenger," reported *Business Week*, "can without any cost arrange with any ticket agency in the United States to have a drive-yourself car waiting beside the station when he gets off the train at Ottumwa, Amarillo, or any other of 150 major cities in 30 states."[35]

Railroad advertising had long touted the automobile as an adjunct to pleasure travel by rail. A 1915 advertisement in *Travel*, promoting travel on the Southern Railway to western North Carolina, read: "GOLF, TENNIS, HUNTING, RIDING, DRIVING, AUTOMOBILING, over splendid macadam roads through U.S. Forest reserves and CLIMBING to the top of high mountains—'Uncle Sam's Newest Playground.' "[36] Railroad promotional brochures also emphasized motoring. Pictured is the cover of a Chicago, Burlington & Quincy brochure published about 1915 to promote Yellowstone National Park and its access by way of its Cody, Wyoming, station (fig. 3.8).

HENRY FORD AND RAILROADING

No one drew more credit for putting Americans in automobiles and sending them down the "open road" than Henry Ford. Whereas most early automakers built expensive cars for America's elites (the kinds of cars they themselves would want to be seen driving to the country club, critics mused), Ford visualized a well-designed, well-constructed, but relatively inexpensive "peoples' car" for the middling classes. The Model T was his classic effort, conceived and produced based on profoundly visionary ideas concerning production, marketing, and financing. Mass production was achieved through adoption of standardized parts and the perfection of the moving assembly line, the latter through careful time-and-motion study. Mass marketing and a system of self-financing that reduced Ford's reliance on investment bankers were achieved through an innovative dealer-franchise system. Just before World War I, Ford set about purchasing an extensive acreage near the farm where he grew up in what would become Dearborn, a suburb just west of Detroit. There the Ford Motor Company built the world's largest industrial complex: the River Rouge Plant. Needing land owned by the Detroit, Toledo & Ironton Railroad to complete his holdings, Ford in 1921 bought the railroad.

The Detroit, Toledo & Ironton had never been particularly successful. Functioning mainly as a coal carrier from southern Ohio north to the Toledo and Detroit areas (the railroad actually entered those cities on tracks of other lines), it had been in and out of receivership its entire history. Although its ownership would greatly facilitate getting coal to Ford's new factory complex, that was hardly its major benefit. Running some 450 miles north to south, the railroad interchanged freight with several east-west railroad trunk lines connecting New York, Philadelphia, and Baltimore with Chicago and St. Louis. By terminating and originating freight on his own railroad, Ford could bargain down freight rates on those connecting lines and thus substantially lower his costs, both for raw materials coming in and finished product going out of the new plant. Within months of Ford's takeover, *Railway Age* reported that the Detroit, Toledo & Ironton was "receiving a revenue of about 100 per cent above what it would obtain under ordinary conditions for the service it performs, and out of all proportion to that received by its connecting carriers for similar service or similar hauling."[37] But more was at work than bargaining down rates. Henry Ford had also reorganized railroad operations using the same principles he had applied to manufacturing Model T's. In a way, the railroad was fast becoming a mere extension of the Rouge Plant itself.

Ford outlined his operating principles, in a sense challenging the railroad industry to follow suit. First, by buying the railroad, he had rid it of its parasitic stockholders. To Ford's thinking, most stockholders in publically held companies (especially the big investment banks) were only speculators—people just speculating in a stock's value. As such, they were not contributing anything productive to the company. Stock needed to be owned by those who actually worked within a company—not just its managers but its workers as well. Accordingly, Ford had introduced a stock-purchase program for his assembly-line workers, and he did the same for his new railroad employees. Second, Ford set in motion a modernization program charged with bringing maximum efficiency to the railroad, including lighter and more powerful locomotives (much of the railroad would be electrified) and larger, sturdier, but also lighter freight cars. (But his decision to replace wooden with concrete ties on tracks failed. Steam locomotives were literally shaken apart on tracks so provided.) Third, he sought to maximize per-worker productivity by carefully assessing work procedures. In this regard, he introduced to the railroad many of the "time-and-motion" emphases used in setting up his moving assembly lines. Fourth, having increased efficiency, he reduced the number of railroad employees by about a third.

It was the Ford Traffic Department that, beginning in 1918, oversaw the handling of shipments in and out of Ford's earlier Highland Park Plant. Its managers had an "intimate knowledge of all commodities shipped and received" as well as their classifications and the tariffs that governed their movement, *Railway Age* reported. That was especially important for outgoing freight trains since the ICC dictated that, when of different classifications, the lowest-rated freight determined the cost of the whole. Thus Ford perfected the art of mixing freight on box cars; for example, fully assembled cars and racks of car parts were shipped together. The department "sees to it that goods are properly packed, billed and described, and devises loading methods which permit increased utilization of cars, either through new schemes of packing and stowing shipments or through loading different commodities in the same car when mixing is authorized by the classifications."[38]

Some three thousand different parts were required for the Ford Model T. The department tracked the relation of one part to another, the kind of material going into each, and how it was to be manufactured. Thus the average weight per loaded box car shipped by Ford steadily increased. With finished automobiles, for example, car tops might be left unattached, inverted, and filled with seat cushions. Twenty-nine branch plants opened by Ford across the United States assembled cars from the parts so shipped out from Highland Park and later the River Rouge Plant. As for fully assembled Model T's, six touring cars were originally considered the absolute limit for a single railroad box-car. But Highland Park's Traffic Department quickly worked out better schemes. Thus a regulation forty-foot boxcar might contain two touring automobiles, a sedan, a motor truck, a couplet, and a runabout. It was possible "through a well-planned arrangement of the bodies in the car and by attaching the windshields to the walls and bracing the wheels in the corners and on the floor."[39] The Ford Motor Company constantly petitioned the ICC not only for rate changes but also for changes in item classification, only to be turned down.

With the River Rouge Plant in full operation and with a company-owned railroad serving it, the Ford Motor Company not only negotiated increasingly favorable freight rates with other lines, but increasingly the company required of other railroads nighttime interchanges with the Detroit, Toledo & Ironton—and in narrower and narrower windows of time, what today would be called "just-in-time delivery." It was all calculated to reduce the backup of incoming rail cars, but also to minimize in-plant inventories for tax purposes. So also was the railroad able to reduce the per diem rental obligations paid on the freight cars of other lines used in plant delivery. Always the self-promoter, Ford champi-

oned his accomplishments in the popular press. In 1924 one journalist saw nothing but progress when he looked over the Detroit, Toledo & Ironton firsthand:
"The yards lost their old messiness, the ballast was as neat and straight as Fifth
Avenue curb, the engines were not only clean but shining with an abundance
of nickel plating as though dolled up for an exhibition. And all the men I saw
about were not only in neat, clean overalls, but also they each seemed to have
something definite to do and to be doing. The old poorhouse atmosphere had
entirely gone." Ford's management principles, he said, included: (1) do the job
in the most direct fashion, (2) pay every man well, and (3) put all machinery in
the best possible condition.[40]

It was doubtful, however, that Henry Ford's methods would work equally
well with other railroads. As the editors of *Railway Age* argued from the beginning, any success he might have at railroading would accrue directly from the
fact that he owned one of the nation's largest and most successful of industrial
enterprises, one that generated a tremendous amount of railroad business. His
railroad was favored just like the railroad subsidiaries of the nation's large steel-
makers. Those lines had "been very well managed, but their prosperity has been
largely due to the fact that they have been owned by an industrial concern which
controls a vast amount of freight."[41] Ford also benefited by being an automaker.
The automobile industry was not government regulated. It had yet to come under strict union labor rules. It enjoyed government largess through subsidies to
highway building. Ford's was the best of all worlds.

Of course, Henry Ford had to be listened to. The Ford Motor Company's
successes had not made Ford merely an American hero; for better or for worse,
he was a public figure who held stubborn opinions about nearly everything, including, of course, what ailed the nation's railroads and what they needed to do
about it. Ford's experiments with trains, however, ended within a decade. In 1929
he sold his railroad to the Pennsylvania Railroad, with the Detroit, Toledo &
Ironton becoming yet another of the Pennsylvania's many subsidiaries. Ford had
become disgusted by the ICC's constant interference. And he was frustrated by
his inability to change its practices. Yet, many of the nation's railroads did come
to adopt various changes Ford advocated. They introduced new accounting and
car inventorying practices, some of them, of course, mandated by the ICC. They
modernized physical plant, adopting heavier rails on main lines, more powerful locomotives, and lighter rolling stock, all of which enabled longer freight
trains. They instituted better interchange procedures despite ICC restrictions on

pooling. In the passenger field, they modernized coaches, introducing such innovations as air conditioning and sleeping compartments. We discuss the streamlined passenger trains of the 1930s in chapter 7.

CONTINUED LOBBYING FOR CHANGE

In 1930 Daniel Willard, president of the Baltimore & Ohio, summarized the railroad industry's difficulties in an article for *Railway Age* entitled "Taxpayers, As Well As Railways, Are Victimized." Although he was in essence "preaching to choir," his message reflected what the railroads were telling the American public and arguing before Congress. With some 250,000 miles of main line track, representing an investment somewhere in the range of $26 billion, railroads, he wrote, were "performing a freight service each year equivalent to the moving of 3,700 tons of freight one mile for each man, woman and child constituting our population, and a passenger service equal to transporting each one our present population 250 miles a year." The railroads as a whole paid taxes amounting to about $400 million a year. With some 1.5 million employees, railroad payrolls totaled over $2.7 billion a year. Their combined payments for interest on mortgage indebtedness amounted to some $600 million annually, and in 1929 they paid out over $500 million in stock dividends to some 820,000 stockholders.[42] Thus railroads remained critical players in the American economy so imperiled by ongoing economic depression. But under existing governmental policies, the railroad industry's ability to serve the nation could not be guaranteed. Whereas two decades earlier the railroads had carried some 90 percent of the nation's freight, the percentage was now down to 75 percent. Other transportation modes were ascendant. Outright government subsidies and negligent taxing policies were at fault, privileging, as they did, transport by water and air but most especially by highway. The railroads, of course, paid for the maintenance of their own rights-of-way, and provided and maintained their own terminal facilities. Barge lines did not pay for river improvement. The airlines did not pay for airports. On the other hand, out of every dollar earned, the railroads paid twelve cents (or 14 percent of gross revenue) on maintaining tracks and other infrastructure upon which they then paid property taxes in addition.

The nation's inland waterways, excluding the Great Lakes, had an aggregate navigable length of some twenty-seven thousand miles, only 45 percent of which was more than six feet in depth. Nearly half of that mileage was in the Mississippi-

Ohio River system, which the U.S. Army Corps of Engineers was working to bring to a nine-foot depth. Total government expenditure between 1924 and 1929 on Mississippi and Ohio navigational improvements (as opposed to flood control) was over $449 million.[43] In addition, the federal government operated a barge line (the Inland Waterways Corporation, which later became the Federal Barge Lines) in direct competition with the railroads. As another article in *Railway Age* trumpeted, "Inland Waters' Transport Cost Exceed Rail by 50 Per Cent." A sub-head added: "Government subsidy, allowing low rates, only reason for success of slow and inefficient agency."[44] The airlines, besides being provided with municipal airports at major cities and navigation beacons across the country, were heavily subsidized in their operations through U.S. Mail contracts. Mail on the railroads produced only 1.8 percent of total railroad revenue in 1929.[45] However, domestic mail was being diverted in ever increasing amounts (from 700,000 pounds in 1926 to more than 9 million pounds in 1931) to the airlines. Mail contracts flew the planes, thereby enabling the airlines to add passenger service. Air travelers increased from some fifty-seven hundred in 1926 to over half a million in 1931.[46]

Again, it was mainly highways that were the problem. It was bad enough that automobile registrations increased from 9 million in 1921 to more than 22 million in 1931 (or almost two and a half times). However, there were in addition some ninety-eight thousand motor buses in the United States, about a third of them operating between cities in direct competition with passenger trains. The number of bus passengers carried in each of the previous five years had increased from about 1 million to 3.5 million, while railroad passengers had decreased some 30 percent.[47] A cartoon in *Railway Age* originally published in the *Chicago Tribune* told the story: railroads paid their own way, but highway motor carriers, not only the bus lines but also the truckers, did not (fig. 3.11). Although the railroads were willing to concede that some types of freight haulage—for example, the pickup and delivery of packages associated with less-than-carload shipping—might best be done by truck, they still resented the loss of that business. Less-than-carload freight traffic stood at some 53 million tons in 1926, but slipped to 39 million in 1929, and 23 million in 1931.[48] By the 1930s, it was generally conceded that the railroads were best at carrying bulk commodities by the carload over long distances. However, it was apparent that truckers were well on their way to cutting in on long-distance bulk carrying as well.

The bus and truck companies, as public carriers, were relatively unregulated by the states, and not at all at the federal level. Interstate Commerce Commission

Courtesy of Chicago Tribune

There Seems to Be Something in What He Says

FIGURE 3.11. "There Seems to be Something in What He Says."
"Railways Handicapped in Meeting Highway Competition,"
Railway Age, n.s., 93 (Dec. 3, 1932): 811.

oversight of the railroads was quite a different matter. "Railroads cannot build an additional mile of track without securing governmental authority, but interstate bus operators and most truck operators may start or extend operations when and where they please," *Railway Age* argued. "The railroads cannot discontinue unprofitable railroad lines without securing permission from regulatory authorities, but unregulated motor vehicle operators may discontinue service overnight, without notice." Additionally, "the railroads must publish and adhere to their rates, they must charge rates which are fair and reasonable and not discriminatory, and they cannot change their rates except on due notice. On the other hand, with few exceptions, motor vehicle operators may charge what they please, they may be discriminatory in their rates, they may change their rates at will, and they may go to any extreme in cutting rates to get business."[49] Indeed, discrimination between shippers and localities, which the railroads had to scrupulously avoid, was practiced daily by highway carriers. Railroad wage and other labor disputes were necessarily arbitrated by an ICC board. But not so for motor carriers. Thus, working conditions were poor, with truck drivers and bus drivers tending to work long hours (often well beyond basic rules of safety).

In 1900 the automobile was no threat to the railroads. Cars were very primitive and, accordingly, quite unreliable. Besides, there were few places to drive comfortably, save on improved small-town and city streets. Rural roads had been neglected since the 1830s when steamboats on the Great Lakes and the nation's inland rivers, and then the railroads, came to the fore. The nation had lost interest in rural roads except insofar as they enabled movement of farm produce over short distances, mainly to railroad sidings. The nation's rural roads were in very bad shape and thus were largely impassable for early automobiles, particularly in inclement weather. However, once road improvements began in earnest about 1910, and especially once Congress threw itself fully behind the Good Roads Movement with the Highway Act of 1920, the situation changed; intercity motoring over long distances not only became possible but highly popular. Additionally, the inland waterways were reborn through federal subsidization, and the newly created airlines were federally subsidized as well. And it all came at the very time that the ICC undertook tougher railroad regulation. Any monopolistic (or potentially monopolistic) position that any given railroad might have enjoyed, or America's railroads altogether for that matter, evaporated in the face of both rising new modes of transport and continuing ICC regulation focused on the railroads alone. But would the public take notice? Would Congress be excited to act in the railroads' interests? The answer was yes. But it would require a deepening depression and a new political administration and Congress in Washington. Looming on the horizon was the Motor Carrier Act of 1935.

Chapter 4

RAILROAD CROSSINGS

The rapid evolution of motoring in America put motor vehicles and railroad trains in direct physical contact as well as economic competition. By 1910 electric interurban lines in and along rural roads placed cars, trucks, and buses in close proximity to traction cars. But it was at railroad crossings that the greatest dangers seemed to lie. Instances in which motorists tried unsuccessfully to race trains to crossings were all too frequent, and when these incidents were readily reported in the newspapers, it created in the American mind, a kind of disconnect between railroading and motoring. Also problematic was the congestion caused by long freight trains passing through intersections, as well as the shuffling of cars in railroad yards that could block crossings not just for minutes but sometimes for hours. Ways and means of improving crossings using warning signals, gates, and other devices came to the fore, as did ways and means of eliminating crossings altogether with, for example, highway overpasses and underpasses, track relocation, and the building of bypass streets and highways.

THE EARLY YEARS

Crossing railroad tracks was problematic even before the appearance of numerous automotive vehicles. In earlier days, runaway horses posed particular dangers.[1] Most of the accidents happened in open country.[2] In the nineteenth century, when horseback riding and horse-drawn vehicles were the means of transportation, riders and drivers usually stayed well back from railroad crossings with passing trains and knew ahead of time where the crossings would be. Their knowledge was drawn from the fact that the areas of travel were fairly well circumscribed (about twenty-five miles from home)—that is, people were familiar with their home areas. Familiarity and speed changed with the coming of the automobile and other motorized vehicles. *Railway Age Gazette* reported this consensus in the railroading world nearly two decades into the twentieth century.[3] Increased railroad traffic admittedly contributed as well.[4]

Where crossing signals had been built beside ordinary crossings throughout the nation by 1900, what people saw beside the crossing was a post with crossed boards atop, the "crossbuck," bearing the simple declaration "Railroad Crossing" (fig. 4.1). This was adequate in the era of pedestrian, horse, and horse-drawn traffic, but things especially began to change as automobiles entered the scene. The death toll due to railroad accidents, whatever the circumstances, seemed staggering to an editorialist writing in 1905 in the *World's Work:* "War becomes mild when compared with the human havoc wrought by our railroads. . . . the killing and maiming on our railroads goes on year after year, every year's death record usually surpassing its predecessor, the increase in casualties keeping right abreast of our vaunted American progress."[5] What to do about the fact that many more Americans died in railroad accidents than anywhere else in the world and one journalist's claim that it was because of indifference on the part of both the railroad corporations and the public?[6]

Automobility introduced at crossings new considerations aimed at uniformity and heightened alertness over larger areas. Taking 1900 as a starting point, the time when these vehicles' appearance began significantly to cause multiple crossing problems, the *Railroad Gazette* largely supported a new law in Rhode Island, where the state railroad commissioner stipulated taking some precaution—a gate, a flagman, or other "appliance"—at any crossing where it seemed reasonable,. The problem involved not only the cost that the railroads had to pay but the invasion of local authority. The railroad trade journals thought it important that the state law did away with the variations in local decision making, with some localities requiring protective measures and others not requiring them. A single state commissioner would make a judgment from the many cases across Rhode Island and not leave it as an erratic exercise from place to place.[7] Deaths

FIGURE 4.1 A typical crossing designation near Collinsville, Illinois. Authors' photograph.

in two accidents the late 1890s in New York induced an educational discussion in *Railway Age* that elaborated on the state law about crossings at grade level to the effect that railroads must maintain boards, easily seen, on each side of crossings on public roads or streets, painted in capital letters, nine inches in height and "suitable" width, bearing the following words: "Railroad crossing—Lookout for cars . . ." The deciding New York court held that such signs be posted in cities and villages only if street officials required it. New York apparently was behind most states in requiring less expensive signboards along the roadside; the state stipulated maintenance only of "more or less conspicuous" signs and not "across," that is, spanning the roads.[8] Meanwhile, in Chicago, the Illinois Supreme Court gave the right to railroad companies to elevate their rails at crossings and increase their tracks at the crossing regardless of any deficit inflicted on adjacent property owners; this was, so *Railway Age* believed, the proper response to change that had justified expenses levied upon railroads to build crossings, viaducts, or raise tracks for public protection.[9] Chicago undertook a vast program of track elevation via a series of ordinances, the earliest dating from 1892, even before automotive vehicles gave reason to accelerate the program in the early twentieth century.[10] Where the U.S. House of Representatives functioned as local government—that is, in in Washington, D.C.—the Pennsylvania and the Baltimore & Ohio railroads were required to abolish grade crossings, along with altering routes into the city and building a new station, at the considerable cost of at least $1.5 million; meanwhile, the U.S. Senate permitted the Pennsylvania line to build elements of a new freight yard.[11]

Promoting railroads' considerations, the *Railroad Gazette* carried an executive of Pennsylvania Lines West's contention that public convenience, not public safety as calculated at the time, should determine grade crossings. The argument was as follows: All the required safety technologies, such as automatic couplers, heaped huge expenses upon railroads, even when shared with states; and when the actual number of accidents at grade crossings was surveyed, it was calculable that the new grade crossings were not needed and very large existing ones should be abandoned. Track elevation in residential areas was sensible but not in industrial areas. Danger at railroad crossings had been exaggerated.[12]

Piquing response from the trade journal about an earlier assertion, a letter to the editor stated that a bell at a grade crossing was not a luxury but an attendant at the crossing would be one. The trade journal granted both points hesitantly: "The question turns on the degree of reliability. Crossing bells, generally, have no good reputation. Many competent observers believe that they do more harm than good, as they ring when no train is approaching and fail to ring when one is approaching."[13] Later, another respondent defended the reliability of crossing bells and noted that they certainly were cheaper than charges against railroads for unprotected crossings.[14] Another railroad official, however, believed that automatic crossing signals had improved, performed very well, and functioned around the clock. Those alarms triggered by the approach of moving trains were surpassed; automatic alarms were especially good in instances when trains broke apart, leaving behind a separate section to retrigger the alarm at the crossing.[15] But, it was impossible to force upon the public changes in advance of their thinking. Even in the case of Massachusetts, whose railroad crossing litigations and laws the *Railroad Gazette* deemed most commendable, the trade journal reasoned that laws could not be enacted for "any higher standard of conduct than public opinion requires." Perhaps inclined to grant a reprieve from uniform standards, the trade journal elaborated on a Massachusetts law that would have challenged state road commissioners: "The spirit of the law was believed by the legislature to be too rigid, and the people along the line of the road were allowed to have their road, with its dangerous crossings, because it was believed to be better so than to have no railroad at all."[16]

Other factors included aldermen who often voted for provisions regarding crossings because, without expense to the municipality, they were perceived to add a "metropolitan appearance." In rural areas only a few farmers were sufficient to support crossing changes, and they were passed since it cost their towns nothing. It should be understood that the increased danger at crossings was due to increased highway rather than railroad travel. It appeared, too, that people were becoming less careful at crossings. The *Gazette* recommended penalties for careless approaches to railroad crossings.[17] For railroads failing to comply with the local ordinance of Joliet, Illinois, to elevate their tracks in 1903, the mayor threatened to build stone walls to block the crossings or block the tracks. It seemed, however, that the city could use its legal authority to fine the railroads not in compliance with the city's legal directive to elevate the tracks. In fact, the city could not legally block trains because that would involve interstate commerce.[18]

How to finance abolishing grade crossings in city streets? That question grew more serious, especially because of high-speed, long-distance electric railways in highly populated areas.[19] However, with increased traffic over rail crossings resulting from the growing use of automotive transportation, the issue of new public highway crossings, as well as an end to existing ones, became foremost. The chairman of the Alabama Railroad Commission, at a meeting in 1902, glowingly articulated railroads' resistance: elimination should only occur by "fair and lawful means," southern railroads having invested their capital "in the upbuilding and civilization" in a way "as marvelous as Aladdin's fairest creation."[20] Indianapolis began wrangling with the costs of track elevation in 1903, especially because of a large sewer required over Pogue's Run, a creek running through half of the city's southeast side. But, as *Railway Age* reported, the city was about to agree with the railroad for 20 percent of the expense.[21] The state supreme court had declared unconstitutional the city's legislation to force costs upon the railroads. The court held that it required state legislation, and a bill was introduced to split costs along the lines of Indianapolis's pending 80–20 settlement.[22]

The *Railway Age* began to engage technological questions regarding crossings. The costly payment of men posted at crossings to raise and lower gates, especially if they were present each day (annually $600 only on day) and night (annually $1,200 around the clock), or danger signals using a red lamp at night and a flag in the day, were almost as expensive as a crossing guard. Crossing signals required $200 with $20 maintenance per signal. The trade journal advised that the desired signal should be posted fourteen hundred to fifteen hundred feet before the crossing, ring for twenty seconds before any train approached the crossing, stop ringing after the trains passed, keep ringing if another train approached after the first train passed, and be an electric block signal. If more than twenty seconds were given, it had been shown that travelers would try to cross before the train.[23] *Railway Age Gazette* publicized a product known as the Wilson Railway Gate because its metal parts lasted longer than current products; the journal also reported European examples of electric crossing gates (fig. 4.2).[24] Motorcar service came to be included among heralded innovations.[25]

The magnitude of programs to do away with crossings at grade—either elevating them above or depressing them below grade—could be considerable. One of the most far-reaching programs, besides that in Chicago mentioned above, was that of the Pennsylvania Railroad. A total of 994 grade crossings on the heaviest-trafficked lines between New York and Washington and Philadelphia and Pittsburgh were eliminated between 1902 and 1908, only a little more than half of those in the plan. Of 101 grade crossings between Altoona and Harris-

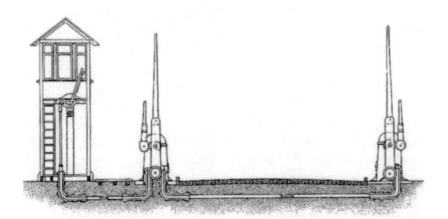

FIGURE 4.2. A sectional view of the Wilson Railway Gate on site. "Wilson Railway Gate," *Railway Age Gazette* 55 (March 18, 1910): 810.

burg, 50 had been taken care of during the same period. Elevated or depressed grades in Philadelphia were especially important safety enhancers because traffic was so dense.[26] Grade crossings in Buffalo, New York, were not so successfully handled. Any privilege granted a railroad that did not shunt traffic to his city, said the mayor of Buffalo, should be revoked. *Railroad Age Gazette* contended that this mayor's response was fair because he had so many years of experience in serving on special commissions.[27] Safety should not be the only consideration, in the thinking of this journal's editor.

The subject of what caused accidents at crossings endured, technological improvements notwithstanding. The *Railroad Gazette* restated the principle known colloquially as the "Stop, Look and Listen Rule." In Pennsylvania, with its lively exchange of opinions and extensive grade alterations underway, it had firm legal standing; simply put, if a traveler could not get a view from inside the vehicle carrying him or her, state law required that the person get out and walk to the crossing. Resulting problems, including death, did not absolve the victim of responsibility and nullify any claims laid upon the railroad.[28] To understand the near hostility complicating this issue in those early days of the century, one need only consider the antipathy that the chief executive of the Long Island line directed toward the public's role in the matter. Declaring that this railroad "has done as much within the last ten years as any road in the county, in the elimination of grade crossings at grade, and, in general, to educate the public in the dangers involved," the chief executive pronounced that the line would fulfill its legal responsibilities but "will not assume the position of public guardian and protector of every thoughtless user of the highway."[29] The *World's Work,* widely read, promoted the editorial stance of the *Railroad Gazette* that another human contributor to accidents, labor unions, were "the most important single cause." Lesser factors were the lack of proper crossings, fences, guards, watchmen, "excessive speed," rabid competition between main lines, and "the innate recklessness of our people."[30]

Planning in the city and port of New York was very complex. The chief executive of the Amsterdam Corporation advocated more effective consolidation in the transport of freight between carrier and consumer than had occurred in the early years of the twentieth century. Although passenger transportation had improved in the forty years consideration was given, especially to freight in the city and port, the carrier-to-consumer connection languished. He favored, among other measures, a freight subway that would alleviate the amount of vehicular traffic on the streets, as well as cheapen freight charges. Failure, he warned, risked the city's national first rank as a port; Chicago's problems, owing to a too-small subway, traffic crossings at grade, and congested railroad yards, should be a lesson to learn.[31]

In the second decade of the twentieth century, the annual report of a New York public service commission's district can be taken as an agenda regarding crossings. Having appreciated the onset of the automobile age because of increased automobile usage and a growing population, New York understood that it had fallen behind other states in handling this problem despite its ambition to be foremost in many aspects of public life. There were 9,065 crossings in New York, two-thirds of which lacked protection, and deaths at these intersections continually rose. A large number of highway crossings had been designated for elimination, but the state lacked funds to execute this plan.[32] Inside New York City, work to eliminate crossings was planned for 1913 after the governor vetoed what would have been the first appropriation.[33] In 1916 the National Association of Railway Commissioners discussed a national system of safety measures for crossings eight years after New York's annual report.[34]

Many railroads and their advocates continued assuming a defensive posture about accidents. When, in October 1911, an article about railroad accidents appeared in the *Hampton-Columbia* magazine, *Railway Age Gazette* launched a detailed rebuttal. Its particulars were as follows: American railway accidents were not as bad as had been claimed in comparing them to Europe. Many of those killed in railroad accidents intruded upon railroad property by "stealing rides," or walking on railroad tracks when getting off of railroad equipment, or falling off. Charges that railroads required their drivers to maintain dangerously high speeds in order to deliver freight and passengers on time were groundless because railway unions would have opposed the practice. Labor laws also permitted no more than sixteen hours of continuous service and, in practice, railroad workers labored no longer than ten hours or traveled over distances more than one hundred miles. Automatic devices activated to stop an imminent danger would not prevent many accidents because few were the result of collisions. Contrasts with European railroads also were not fair because American railroads were built in many places where they were intended to attract population, build industry and traffic, and not handle a lot of existing traffic. When growing traffic justified it, improvements could be made and charges that the railroads were not correctly equipped would be justified. When regulations on American railroads also enabled them to invest and earn profits, as did those in Europe, it would be defensible to criticize them for not being better equipped.[35]

Problematic crossings gained no attention in this exchange. In a simultaneous counter to public opinion, the *Railway Age Gazette* charged that the newly arisen Good Roads Movement should factor in the costs of safe crossings and needed to account not only for farmers using improved roads but, even more so, the growing numbers of pleasure-driving motorists.[36] At exactly what locations should the railroad bear maintenance costs because of increasing automobile traffic? Did it include the highways beyond the tracks in the approach to the right of way? Laws varied between states.[37]

Railway Age Gazette offered a somewhat more balanced opinion about railroads and elevated crossings. Repeating the views of a Union Pacific official, the trade journal spoke of the automobile as one of four key improvements that had made contemporary life more convenient, the others being the telephone, the electric light, and the bathtub.[38] Yet, it also reported the findings of the Railroad Commission of California for 1914–15 that not a single fatality involving automotive vehicles at a grade crossing was due to a railroad.[39] The trade journal pushed into wide public view the debate in New Jersey about whether an end to selected grade crossings should be left up to the public utility commission or to the railroads at the rate of one for every mile of road each year. It afforded the president of the Pennsylvania Railroad an opportunity to call for legislation that "will bring the state, the municipalities and the railroad together, insuring the apportionment of the cost in proportion to the equitable interests of all concerned."[40] Finally, the journal granted that "we cannot measure lives in cash" and that Chicago's ongoing program of railroad-track elevation was yielding the desired results. It remained incredible that the public was indifferent about accidents from other causes. The rising popularity of automobiles overrode the Long Island Railroad's efforts to work with those in the new automobile wave to prevent accidents. In today's more objective atmosphere, it seems that, like the public in general, early automobilists wanted to have it both ways: the public was uninterested in shouldering expenses to control accidents, and yet park boards supporting track elevations insisted they should beautify boulevards that increased expense. Auto traffic was somehow ranked higher than the railroads, as in the case of one Chicago boulevard where, despite the hazards in crossing a railroad track, no one wanted the boulevard elevated. And, typically, where expense would be incurred, the public expected the railroads to pay; this the journal repeated like a growing refrain in the public narrative.[41] Still, at the end of 1915, *Railway Age Gazette* re-stated its view that careless drivers produced the overwhelming percentage of crossing accidents.[42] A year later the trade journal reported that data in Illinois showed that automobilists at crossings "exercise much less precaution than driv-ers of horse-drawn vehicles."[43] Further, a year and a half later, the journal could be counted on to report a "comparatively infrequent" accident at a railroad grade crossing where the court laid blame upon the driver.[44]

Systems of protection, including guards and devices, seemed to decline in relevance. *Railway Age Gazette* reported on new crossing devices when they came out (fig. 4.3).[45] It twice stated support for a device signaling speed reduction be-fore crossings and added that a curve in the road before the crossing would be a functional complement. However, the writer's excitement for such measures lapsed, at end admitting that reckless drivers would be so nonetheless.[46] A bill under consideration in the New York legislature shortly thereafter requiring au-tomobiles to be stopped before proceeding over a railroad crossing was neces-sary because, regarding motorists, "the only workable rule is a severe one."[47] New Jersey's legislature played a public trick on railroads, or so *Railway Age Gazette* opined, by ordering railroads to maintain gates, flagmen, or automatic bells at every crossing of a certain class of highways only because the previous law re-quiring the elimination of all grade crossings would likely have been challenged and annulled.[48] The trade magazine reported only in a matter-of-fact manner on New Jersey's "important addition" to the standing state law that railroads had within one year to have railroads provide safety gates, flagmen, electric bells, or other systems of alarm or protection.[49] In the case of an automobile that ran into a train on a very dark night, a New Hampshire court decided that no damages were owed to the automobilist unless it could be proven that the only cause was the railroad's failure to provide a signal.[50] When technological warnings were not fully posted on the Long Island Railroad by mid-1915, it was deemed satis-factory to post highway signs—bigger than in the past, but still signs, appealing to drivers' sense of caution. As one sign read:

THIS SIGN MAY SAVE YOUR LIFE TODAY.
All The Precautions in The World Will Not Save The
Lives Of Those Who Drive Automobiles Recklessly
Over Railroad Crossings.
When Approaching A Crossing Please Stop, Look, & Listen
We Are Doing Our Part. Won't You Do Yours?
        LONG ISLAND RAILROAD.[51]

The railroad combined this with a newspaper advertising campaign stating the number of collisions at crossings in which reckless driving was at fault.[52] A railroad man in Texas, claiming to speak for automobilists, wanted uniform behavior among flagmen at crossings, and the *Railway Age Gazette* editorialized for disks bearing the word "STOP," rather than flags, to aid flagmen.[53] In figure 4.4, the "Big Four" Depot at Terre Haute, Indiana, is shown in about 1910; a crossing guard stands next to the station leaning on his "stop sign." Note that the crossing is otherwise unprotected. Legislation about safety at crossings, the *Railway Age Gazette* predicted, would soon extend nationwide from its core in New England, California, and Chicago.[54]

FIGURE 4.3. Signal and alarm fitted with a Chicago electric flag. "A Highway Crossing Alarm With Four Indications," *Railway Age Gazette* 58 (March 19, 1915): 677.

Alternatives in places of numerous crossings turned to huge engineering projects to build solutions. The Delaware, Lackawana & Western, for example, undertook a system of bridges and viaducts—including what was then the world's largest

FIGURE 4.4. Crossing and crossing guard at "Big Four" depot at Terre Haute, Indiana, circa 1910. Postcard, authors' collection.

concrete bridge—at a cost of $12 million to replace lines.[55] The segment through Orange, New Jersey, alone took two years; it eliminated twenty-six grade crossings, replaced two-track portions of the line with three tracks on an improved alignment, and enlarged facilities for freight and passengers.[56] In St. Louis, both the Missouri Pacific and the St. Louis & San Francisco shared with the city the cost of $830,000 over one and a half miles.[57]

Railroad sympathizers continued interlarding the discussion throughout the nation with the reminder that the public must still bear some responsibility for safety at crossings and that this could sting because it could be translated into money. On a country road near Danville, Illinois, in mid-1916, a driver was killed at a crossing despite crossing signals at work. The wife of the engineer who died in the wreck filed suit against the estate of the deceased driver, and the railroad filed suit as well for the damaged train. *Railway Age* dilated on the tragedy in print.[58] *Railway Age Gazette* closed a long discussion about the comparative merits of audible versus visual crossing signals with the admonition that the public must be taught to be cautious for it loses confidence in a safety system when it fails because "the confidence of travelers is blind, and often ignorant and wrong headed."[59] *Railway Age* continued emphasizing that drivers should take more precautions.[60]

Railway Age judged that, overall, the final few years of the nineteenth century and almost all of the twentieth century's first two decades were filled with vexing issues related to "the grade crossing problem."[61] The dawn of the automobile age had made it so. Progress was being made, however. The journal reported Connecticut's adoption of a system of uniform signs to be adopted statewide.[62] Cautionary highway crossing signs embodied the most progress in this period, according to *Railway Age Gazette;* laws in eight states required them by late 1917. These supplanted the earlier remedy of eliminating grade crossings because it had proven too expensive.[63] Concrete began to be appreciated as a superior substitute for wood used on highway crossings.[64] Individuals workers, not only executives, systems, and engineering, shared in signs of promise. Implying that women were welcome into the ranks of those who waved cautionary flags at crossings, *Railway Age Gazette* and the *Train Dispatchers Bulletin* chronicled the case of Marie Travers in Cincinnati, Ohio, who entered the ranks and served so that men could enter military service in World War I. The Baltimore & Ohio Southwestern had initiated a program to recruit "flagwomen," believing they were no less capable than men in this work.[65]

Yet puzzles still plagued crossings. Chicago, for example, got rid of 963 grade crossings during the twenty-seven years the program was in effect. Deaths dropped from one hundred in 1906 to fifty-nine in 1913 but began rising again. In 1916 they increased almost 50 percent.[66] Would speed bumps in front of crossings or other devices to restrict speed upon approaching crossings save lives and spare injuries? The Nashville, Chattanooga & St. Louis Railroad thought they would. Or, would they merely shift responsibility from the railroad to the driver in the event of an accident, as they had in one case in Memphis where the court ruled that the bump had given sufficient warning?[67] Just before the start of the period traced above, grade crossings had been the scene of 675 deaths in in 1898, but 2,500 deaths were anticipated five years into the next period.[68] Solutions seemed more urgent.

THE 1920s-1940s

Pausing in 1925 to comment on the decline over the previous year of automobile accidents at grade crossings, the editorial staff of *Railway Age* still took the opportunity to expand on what it considered the irresponsible remarks of automobile manufacturers. One automotive executive said that railroads should spend more money on improved operation, especially at grade crossings, rather than on legal fees to defend against damage suits and on the repair and replacement of equipment ruined in collisions; in so doing, they could perform work "beneficial to humanity and of permanent value to themselves." Little did the manufacturers understand how much improvements cost, the journal argued. If the most effective means were instituted—namely, eliminating grade crossings—the cost at the end of 1923 would have been $19.5 billion.[69] Each side still had a strong penchant for blaming the other.

A presumably more objective spokesman, Connecticut's chief engineer of public utilities, referred in 1928 to the "mania for motor car speed" and went on to itemize remedies for accidents at grade crossings: standardizing signs at grade crossings, decreasing the number of unprotected or part-time protected crossings, using horizontal flashing-light signals only to indicate a train's approach, educating the public about what to do at these locations, urging punishment for reckless or careless driving over a grade crossing, and enacting legislation to prevent construction of more grade crossings.[70] Techniques for on-site control, thus, still earned approval but the appeal for public education did too; for example, *Railway Age*'s article of 1925 cited above found remedies much like Connecticut's chief engineer but said of public education that it was "the most important of all."[71] In summers, the Safety Section of the American Railway Association conducted a "Careful Crossing Campaign," which in 1929 included posters (542,645 printed) and booklets (613,850 printed). The American Legion assisted the effort by displaying the motion picture "Look-Listen."[72] In 1934, the Erie's program of "the last few years" to reduce highway crossing accidents gained attention in *Railway Age* and included "educational work" such as talks at schools, service clubs, and newspaper publicity.[73]

Numerous educational programs were developed, beginning in the early 1920s. In combination with a local newspaper campaign, the electric street railway in Milwaukee mounted a sign on a work car with safety warnings to automobilists.[74] At least twelve years earlier than Erie's program, Chicago witnessed one of the most extensive educational campaigns, including posters, stickers, and the participation of various clubs (including the Boy Scouts of America) and public and parochial schools. Publicity in the press was added. An area extending to eight hundred miles around Chicago was reached. A special pamphlet urged the cooperation of an estimated 550,000 school children.[75] In 1922 the Westinghouse radio company in Chicago, Pittsburgh, Newark, New Jersey, and Springfield, Massachusetts, broadcast the results of the Pennsylvania Railroad's "Safety First" campaign, which *Railway Age* considered one of the best in the nation,[76] although only four months later Pennsylvania reported an increase in deaths at crossings.[77] That same year *Railway Age* reported that the National Bureau of Casualty and Security Underwriters had launched the first effort by an insurance company to warn about the circumstances of the increasing number of automobile accidents.[78] The New York Central Railroad circulated a prepared speech, "Cross Crossings Cautiously," for various public audiences.[79] This produced an estimated two miles of newspaper print in leading city newspapers.[80] The *World's Work* in 1923 carried columnist Floyd W. Parson's article about the monetary savings that would result by teaching safety in public schools, presumably with railroad crossings in mind among other accident sites.[81]

New technological warning devices, engineering programs, and legal programs continued, along with the new educational component. It never was taken for a cure-all. Keeping statistics specifically about accidents at railroad crossings began with the rise of automobility, the first national statistics on railroad accidents having begun in 1888. Despite the decline in 1924 to ten fatalities from automobiles at grade crossings from twenty-two in 1917, *Railway Age* classified the rate as "appalling."[82] Statistics could inform educational programs, but faith in devices and engineering programs was paramount.

Engineering included elevating railroad tracks. In 1921 in Patterson, New Jersey, this took the form of a legal demand from the state's public utility commission, backed up in a case before the U.S. Supreme Court. *Railway Age* complained that this was an unfair expense to the railroad.[83] Devices included illuminated highway crossing signs, like those of the Northern Pacific Railroad publicized in *Railway Age*.[84] Opening new grade crossings fell from favor among railroads.[85] For urban areas, railroads seemed less concerned about remedies for accidents at crossings than in rural areas, where the incidence of accidents was higher, instead proposing the closure of less-trafficked urban streets. Zoning did not appear helpful because railroads and cities were codependent upon business locations that had been extensively developed before the concern about railroad-crossing collisions arose.[86]

Engineering solutions included materials, especially the concrete widely employed for grade separations that tunneled automobiles through arched and flat-topped entries while having the railroads pass overhead.[87] The American Railway Association's Committee on Signs, Fences and Crossings considered uniformly adopting concrete slab crossings, concrete plank crossings, concrete fence posts, and the proper proportions of materials to mix concrete.[88] Monumental constructions included a 784-foot-long concrete arch viaduct, 105.5 feet high from the bottom of its footings, at Sidney, Ohio, for the Cleveland, Cincinnati, Chicago & St. Louis line.[89] "Island" or "fool-proof" construction called for an obstruction in the approach to a crossing that would naturally compel drivers to slow up.[90] Paved street crossings of railroads remained highly variable from place to place.[91]

Equipment included automatic crossing signals in place of gates with flagmen.[92] But advice was given simultaneously about how best to coordinate crossing signs with either gates or flagmen; the latter had not gone out of use by 1922.[93] Wig-wag signs, so called because of their pendulum-like motion, had gone out of favor in all but the West by 1924.[94] Flashing signals replaced crossing gates in some places, with some using three colors like automobile traffic signals.[95] Pictured at Bureau Junction, Illinois, is a crossing of the Rock Island Railroad with an electric crossing signal and crossing gate on a very busy intersection; its use dates back to the 1930s (fig. 4.5). As late as 1935, however, the replacement of flagmen with automatic crossing gates still gained the *Railway Age*'s attention.[96]

FIGURE 4.5. In the 1970s, the Chicago, Rock Island & Pacific Railroad languished in insolvency, with little maintenance done on its main line. This photograph at Bureau Junction, Illinois, in 1973, is a record of an antiquated crossing signal. It shows as well the surviving ambiance of the early twentieth century main street. Such can still be encountered, especially in the Midwest and the Northeast. Author's photograph.

One less frequently applied solution involved legally required practices among adjacent states.[97] Five states, by 1923, required automobilists to stop before a grade crossing.[98] This was but one of the streams feeding proposals for standardizing laws that would bring widely and easily understood coherence to different statewide steps underway in the 1920s.[99] The Committee on Construction and Engineering of the National Conference on Street and Highway and Highway Safety moved in that direction at the end of 1924.[100]

Advocacy became stronger for cooperation between representatives of automobilists and railroad interests. Charles E. Hill of the New York Central Railroad spoke in its favor in 1924 and congratulated the National Association of Railroad Utility Commissioners, the American Automobile Association, the National Automobile Chamber of Commerce, the National Safety Council, the railroads, state highway departments, and other unnamed but interested organizations.[101] In a unique case, a bus line headquartered in Columbus, Ohio, found it financially

FIGURE 4.6. A modernist design of a concrete casement for a new railroad crossing at Bellevue, Ohio. Postcard, authors' collection.

feasible to erect lighthouses at twenty-five railroad crossings in its six hundred miles of routes, but this was so only because the number of trains had declined and been replaced by buses.[102]

How to pay the cost for changing crossings, of course, was an enduring matter. An executive of New York's Transit Commission recommended bond issues but also borrowing from the state at lower interest rates.[103] Railroads were upset with the lack of unified standards because of the approximately 150,000 political jurisdictions nationwide that set the standards.[104] With the federally inspired New Deal, jurisdictions were granted their own wishes in railroad-crossing projects.[105]

Blame for which side—railroads or automotive vehicles—caused the accidents droned on into the 1930s.[106] In 1933 *Railway Age* picked out a series of oil truck accidents at crossings to underscore that these mishaps were due to the truck drivers.[107] In an effort to find agreement between both sides, the journal printed lengthy statements from spokesmen of the two viewpoints.[108] It also published the call for cooperation by Thomas H. MacDonald, the chief of the U.S. Bureau of Public Roads, who said that railroads were an asset, not "a monopoly to be curbed."[109]

Although many grade crossings were eliminated, new ones were still built, *Railway Age* reported in 1930.[110] A slighter number of crossings, 136, resulted by

1930.[111] Chicago's earlier example of how to reduce grade crossings and substitute grade separations continued to set the standard at the start of the 1930s.[112] Drawing on a study that took six years to complete, Detroit was on the eve of starting work on replacing grade crossings with grade separations in late 1932.[113] New grade separation was taken on in Wisconsin in 1931 as a remedy for unemployment, even though crossing casualties had been an admitted problem for some time.[114] The Chicago, Milwaukee, St. Paul & Pacific Railroad reported plans for an extensive project (2.45 miles of depressed miles of track and 2.55 miles of elevated track) for $4,750,000 in 1934.[115] Meanwhile, Syracuse, New York, undertook a project for the elimination of grade crossings. An ideal overpass to avoid a grade crossing follows in this illustration (fig. 4.6).[116]

The New Deal gave stronger emphasis to grade-crossing solutions; they were included in a general public works appropriation of $4 billion for 1933.[117] A total of $200 million was allotted finally for this purpose to eliminate almost three thousand crossings.[118] Rapidly, at least twenty states began to install or planned to install automatically controlled signals at crossings.[119] A railroad executive called for a united effort between railroad and automotive interests to address the grade-crossing issue, a "national problem."[120] In 1936 the New York Railroad Club provided a stage for spokesmen of the railroads and city and federal highway authorities to discuss a "more equitable division of costs."[121] Availability of funds via the New Deal had invited cooperation, and *Railway Age* foresaw problematic crossings being eliminated through mid-1939 and perhaps extended yearly as part of the regular federal-aid highway funding.[122] The Hayden-Cartwright Act of 1936 had appropriated $50 million for fiscal years 1938 and 1939 to deal with grade crossings, and in 1938 an act approved $20 million for 1940 and $30 million for 1941. They were subsumed colloquially as "federal-aid" projects.[123] New funding for railroad crossings aside, the widely popular *Saturday Evening Post* published an article in 1937 concluding that among most parties involved with crossings—those who financed them, the railroad operators and owners, and motorists—"nobody likes them."[124]

New apparatuses continued to come out in the 1930s. The Chicago & North Western introduced the "Auto-stop," which combined a light and two barriers across the highway with a train's approach.[125] Under the New Deal's Emergency Relief Appropriation Act of 1935, it was installed in some cases. However, older types of warnings were put to work, the fixed-lamp flashing-light signal being the most popular. Though considered obsolete in earlier reports in the railroad trade publications, the wig-wag signal was approved for installments along with its flashing-light accessory.[126] In 1938 *Railway Age* publicized lights installed at crossings in Niagara Falls and especially promoted them in areas of switching and terminals.[127] In 1940, on the eve of the U.S. entry into World War II, *Railway Age* reported harmonious collaboration between the federal and state governments. As of January 31, 1940, 2,060 grade crossings had been eliminated, 380 grade-separation structures had been reconstructed, and 1,115 protective devices installed at the estimated total cost of $192,970,962.[128]

This is not to say that warning systems at crossings were universally brought up to date. Immediately upon the end of World War II, for example, Cary, Indiana enacted a city ordinance to bring the numerous crossings, a total of seventy, up to contemporary standards in that very active railroading town. Eight railroad companies passed through Gary. Several out-of-date devices were to be replaced by standard ones of the time. Mechanical and pneumatic crossing gates replaced hand-operated ones and watchmen. Part of the ordinance stipulated what those crossing the tracks had to do. The *American City* reported that only one person died in a crossing accident in 1948, whereas the previous number annually was five to ten.[129]

THE 1950S TO THE EARLY TWENTY-FIRST CENTURY

John Stilgoe's scholarship about railroad crossings in the previous half-century is well founded. The railroad-crossing dilemmas can be taken metaphorically as signs of danger about "the melding of traditional landscape into something new," as automotive highway traffic hewed to town and suburban living while trains earlier hewed to industrial zones.[130] Americans, however, dwelled more obviously on the literal dangers to lives and property lost in accidents at crossings.

Old realms of disagreement were resumed in public debate. The case of *Shanklin v. Norfolk Southern Railway Company* (1997) in Tennessee essentially concluded from complicated evidence that the plaintiff, the wife of a man in a car killed at a railroad crossing, had questionable grounds for a claim as did the railroad in its defense.[131] States could no longer administer the federally allotted funds for crossings by allowing railroads to supplement their maintenance funds by drawing on federal funds regarding crossings. The railroad worked long and hard to mobilize its best defense using federal safety funds. Railroads paid for signs and signals. States paid for maintenance and, consequently, had to address the costs of inadequate maintenance.[132]

Who would pay? How much damage and life would be lost? Railroads and automobilists in the new automobile age faced questions a century old. Some

250,000 public and private crossings at grade were the scene in 2003 of 2,929 collisions, 329 fatalities, and 1002 injuries.[133] According to a presentation at the 2004 conference of the Transportation Research Forum, it was reckoned that, by the mid-1960s, safety at railroad crossings had "reached crisis proportions." The absolute number of fatalities peaked in the 1930s, but relative to the amount of railroad traffic, fatalities had risen. The railroads' declining popularity, compared to that of the automobile, had made trains no less troublesome in the public's perception. Exchange in the political realm about responsibility and solution yielded the Federal-Aid Rail-Highway Crossing Program in 1973 as part of that year's Federal Highway Act. As a result of the program's Section 130, warning devices at crossings were improved, and advice was given on priorities for different crossing types.[134] In the mid-1990s, "ditch lights" were installed routinely on locomotives and led to a reduction of accidents by one-seventh of the total.[135] Since 1973 two-fifths of the reduction of accidents resulted from reduced drunken driving and better medical response. Gates and/or flashing lights reduced accidents by one-fifth. Better-informed public behavior yielded the benefit of a one-fifth reduction of accidents and one-tenth from closing crossings or consolidating little-used crossings.[136]

Railroads and roads stimulated economic benefits, also helping to build the nation across the continent and foster cultural cohesion. Nonetheless, these huge benefits also brought threats to public safety, with profound consequences for life and property, literally coming together at railroad crossings, where trains and all other surface-transportation modes encountered one another. Automotive transportation multiplied the threats, and because of these crossings, an often bitterly discordant note arose. As combined solutions in engineering, technology, public education, and funding were sought during the first third of the twentieth century, relative harmony arose with federal funding during the Great Depression. The loss of life and property still loomed as threats, and refining established ways to cope with the problem continued through the early twenty-first century. Railroad crossings have embodied one of the most serious collateral subjects in the nation's adjustment to railroads and automotive transportation.

Chapter 5

STREETCARS

Cities required some form or forms of transportation, of course, to develop at all. For centuries, pedestrian, horse-ridden, or horse-drawn forms served well. With the onset of industrialization, including its reliance on ever-improving forms of technology to make better products with ever more profit, steam-powered, electrically driven, and gasoline-powered engines expanded visions and enabled cities to grow from their core. They symbolized progressive urban populations and the capacity for more and better in general. In this chapter we follow the paths of four intracity forms of mechanical transportation as its advocates attempted to increase their uses, if not dominate and operate to the exclusion of the other forms. Because of their centrality in the means of production, we continue to rely heavily on the transportation trade journals for this narrative.

THE ELECTRIC STREETCAR'S RAPID RISE

When cars were new to American city transportation, they were horse cars, that is, horse drawn and not self-driven. Recalling this reminds one of the far-reaching changes experienced in the transportation over city streets in little more than a century. Founded in 1884, the *Electric Railway Journal*'s first name, *Street*

Railway Journal, hints at the transportation transition on urban streets. Horse railways numbered 525 in 300 cities in 1884 with 16,000 four-wheel horse cars plus 100,000 horses and mules. They vastly outnumbered cable cars operating in six cities because they were far less expensive, and electric cars had only existed since 1879.[1] The latter required further electrical equipment to make them practical. The *Electric Railway Journal's* editors claimed simultaneous birth with the first functioning electric streetcar line, which operated in Cleveland. The journal functioned to help develop the electric streetcar industry by exchanging information among its readers and information about inventions and experiments,[2] and its editors wished it to be taken for the industry's history.[3] Not only synchronized with the fledgling industry, it also conveys the mood and feeling of this railroad-based accompaniment in the rising age of automobility. It grew into a weekly publication in 1900.[4]

The journal believed that the electric street-railway system peaked in 1894 and that horse-driven street railways began to decline in 1889. Technological advances being common, is it surprising that the journal believed that changes of the time in electric railways would take over a considerable part of the steam-railway system?[5] St. Louis gained extensive coverage ten years later for its street railway's capacity to handle large volumes of traffic: some 1,151,785 passengers and without big crowds congesting traffic one day of a fair.[6] The six different companies serving Los Angeles and outlying areas also gained the journal's attention in 1904.[7]

In 1904 there appeared the first statement in the *Street Railway Journal* about automobiles on city streets, and it was quite contemptuous, the words of a "social reformer" who believed that automobiles required too much space, pushing aside pedestrians and all other kinds of vehicles.[8] This opinion was largely ignored. Only a year later, the *Literary Digest,* appreciating the problem of the rising competition between trolleys and cars, attempted to stimulate productive public exchange by publishing the thoughts of an Englishman commenting on London, where, he believed, omnibuses should work as feeders of the existing trolley system.[9] "Traffic congestion came on suddenly," it was acknowledged in retrospect in a 1924 article in the *Electric Railway Journal.*[10] Within a decade, traffic congestion had thrust the need to resolve the competition into a foremost practical necessity. Somewhat belatedly, in 1914 the *Street Railway Journal* reported on measures to make urban intersections safer for pedestrians and ease the flow of cars and other vehicles. On Broadway in New York City, painted zones were afforded to pedestrians for standing and in other cities were set aside be-hind iron "standards" or on islands raised above street level. For vehicular traffic, various types of signaling systems were introduced, these obviating the use of policemen to control traffic.[11] The topic of competition for movement within the fixed space of urban streets was now a constant. The journal remarked in December 1914 on the private automobiles, especially in Los Angeles, that transported passengers for a nickel's fare in competition with electric railways (fig. 5.1).[12] In Columbus, Ohio, the Columbus Railway, Power & Light Company's program to bring greater safety onto the streets, not as a public program but as that of the company alone, yielded greater safety as well as financial savings. It was also able to reduce the highest incidence of accidents, that between the trolley cars and all other vehicle types.[13] Even with what would prove to be a mistaken predic-tion about automobile registration—namely, that the registration rate probably would not increase so much by 1920 as it had between 1910 and 1915—electric

FIGURE 5.1. "Broadway on a Busy Day, Los Angeles." Postmarked 1922 from Los Angeles, this postcard shows pedestrians, automobiles, and electric streetcars competing for space. Author's postcard collection.

railways eagerly continued the effort to control accidents with the automotive swarm. This an executive of the Puget Sound Electric Railway attempted to do by endorsing a program principally to regulate automobile drivers.[14] Members well spread across the nation notably multiplied the electric railway section of the National Safety Council in 1915.[15] Chicago's rapid awakening to traffic congestion in the Loop led to an elaborate on-the-ground study of traffic behavior by Chicago Traction; this statistical report came with recommendations based on the assumption that "the few years just preceding 1915 be considered as typical of the normal in Chicago." It concluded that "the only available remedies for this situation appear to be re-routing, more efficient traffic regulation, and subways."[16]

By 1918 congestion appeared to be here to stay. With Dallas, Texas, in mind, an analyst in the *Electric Railway Journal* sketched a multipart remedy for congestion, accidents, and slow traffic; regarding electric trolley lines, it recommended that they not operate in tandem and that they increase maximum speed, grant motormen discretion at crossings with steam railroads, and work only within the downtown business area and not into residential districts.[17] "Street-car and vehicular-traffic requirements must be correlated and adjusted for the good of all the people," the analyst declared, appealing to the nation's democratic moorings.[18] Chicago, six years later, considered a program for rerouting streetcars in the Loop, for although they caused congestion, they and the area they served were so vital to the city that it could not be imagined without them.[19] Pictured is a heavily congested downtown Kansas City street, with transit cars stacked up with automobiles all around—a classic gridlock. Such could be the case in many big-city business districts, especially at rush hour (fig. 5.2). Rerouting not only streetcars but also buses was a key element of the plan finally adopted.[20]

Automobiles' negative capacities were not denied. In general, their drivers were reckless, attentive to other cars but not to pedestrians. In the nation's capital, the possibility was great for collisions because parking was allowed in the center of streets and on both sides of tracks. In the capital they were also one-twentieth as efficient as streetcars. Nonetheless, *Electric Railway Journal* saw them as essential and advocated a system of coordination since streetcars were essential, too.[21] Street railways, the journal commented briefly, were too often accused of poorly sharing public streets without regard for the fact that complaints were the loudest where the original streets were narrow and reduced to "mere lanes" shared with automotive traffic.[22] The American Electric Railway Transportation

FIGURE 5.2. Walnut Street, Kansas City, Missouri, circa 1920. Postcard, authors' collection.

& Traffic Association did not close itself off to the opinion of the police chief of Rochester, New York, who believed that automobile drivers could be more careful and considered the automobile to be a "most useful vehicle."[23] Los Angeles expanded its ninety-foot-by-four-foot safety zones to one hundred feet by five feet, replaced white-painted strips on the street with white cement strips to delineate the zones, and introduced heavier chains around the safety zones where pedestrians were to stand while streetcars came to pick them up. *Electric Railway Journal* summarized: "The plan has proved satisfactory, is speeding up loading of cars and prevents accidents."[24] This was an attempted remedy unlike many others that claimed to be successful.

Different cities handled their urban transportation problems differently. Detroit's remedy for its future transportation problem, based on an anticipated doubling of population by mid-century, was meant to lean heavily on electric railroads in the forms of a subway and elevated railroad system; however, the U.S. entry into World War I forestalled it.[25] The executives of Pittsburgh Railways, noting the city's narrow streets with their steep grades and sharp turns, saw a solution in more automobility. Although automobiles had "done the street railway much injury," they conceded that such vehicles were fast and would not stall traffic should there be "enough street room."[26] Des Moines, Iowa, the state capital, confronted a shocking sequence of events forcing citywide consideration of urban transportation. On August 3, 1921, the city's fifty-year-old trolley line shut down, having gone broke while trying to fund the costs of rehabilitation, extensions, and other requirements under its public franchise. Unable to fund costs with its six-cent fares and unable to get court approval for a rate increase, it ceased operation.[27] Those who did not walk or refused to stay home gave rapid birth to the market for improvised automotive services. Only twenty-five buses had operated before but on the third day after the strike, twice as many buses were in operation.[28] After enduring an eighty-four-day cessation of the traditional electric trolley service, the city's inhabitants voted two to one for new terms to put the electric railway service back into use.[29]

During and after World War I, studies were undertaken in numerous places, and experiments were made in changing rates to finance increased costs. Flat rates had been standard since the industry's birth. Weekly and Sunday passes became the most popular changes. Fares levied according to a system of zones serviced were generally unpopular. In the process of describing the practices prevailing in 1931, *Electric Railway Journal* figuratively stood back in an emotionally detached manner to pronounce its historically derived conclusions: "The usefulness of reviewing failures and successes of past years may be questioned. Conditions change; methods and practices must change with them; but the fundamental policy of giving to the public upon whom we depend for our existence the best available form of service at the lowest reasonable price has remained unshaken throughout the industry's long and interesting career."[30] A positive spirit about the future prevailed. Major financing in 1930 included the Cincinnati Railway's notable issuance of bonds for $18,877,000 to be paid in 1955. An editorialist wrote, "Continuing the good record made in 1929, the electric railway industry came through the past year in excellent shape, in so far as its finances are concerned."[31]

Symptomatic of a service no longer keenly competitive, it was not until 1921, four years after the traction industry was judged to have "passed through the most critical period in its history," that the Federal Electric Railways Commission reported on those conditions. The commission assured those interested that the industry had "met the crisis and is now on the road to convalescence."[32] Of the fifteen root causes underlying the industry's problems, aspects of financial mismanagement were deemed most responsible; automobile and jitney competition was but one.[33] In fact, automobility in its several forms was primary. At the start of the third decade of the twentieth century, it likely comforted the industry's leaders to have faith that "a town that has no electric railway service presents the appearance of a deserted village," but that notion was rapidly proving ill-founded.[34] A realtor speaking to the American Electric Railway Association in 1924 recommended buses as the best transportation option in urban areas but that, when sufficient traffic growth had occurred, streetcars should be introduced; this was how to increase property values in the area served.[35]

Street railways did respond to the competitive challenges by beginning discussions about salesmanship. The *Electric Railway Journal* elaborated on circumstances in Brooklyn, where municipal government had presented greater obstacles than elsewhere, and the several electric railroads in operation worked especially hard to overcome challenges in favor of the passenger, practicing "indirect" merchandising.[36] At the end of 1921, the journal explained the struggles with which street railways wrestled to sell their services and recommended that each company open a special department. It also cautioned not to think poorly of what had been achieved, arguing that street railways would not "build up in a day a rejuvenated business."[37] Acknowledging the contribution of gasoline utility trucks that electric railways already operated and in the future motor buses and freight-carrying motor trucks that electric railways would operate, the journal recommended in 1921 that the electric railways build shop and storage plants suitable to these complimentary transportation forms. Most presciently, the article alerted readers to foresee "the Automotive Age."[38]

By the mid-1920s streetcars were promoted as the best means of mass transit, and urban transportation plans to accommodate them with other transportations types were advanced as a general principle.[39] Detroit resumed its remedy for slowed traffic that had been stillborn because of World War I. Merchants and the street railway convinced the city council to restrict downtown parking. A number of "through streets" were designated; these required all automobiles to stop before entering or crossing them.[40] Upon the street railway's acquisition of bus lines operating on highways, the streetcar service planned for the two means' maximum frequency of service by staggering their services instead of duplicating their stops concurrently.[41] A presentation at the meeting in 1924 at the American Society of Civil Engineers argued effectively, so the journal reckoned, that removing streetcar tracks from congested streets so that motorized transportation could prevail would not remedy congestion because trolleys moved about 80 percent of the urban traffic.[42] People wrongly thought that streetcars inherently slowed and interfered with traffic when, in fact, moving in the stream of mixed vehicular types was the cause of the sluggishness. The executive of the Eastern Massachusetts Street Railway in Boston, who accepted this view, advocated changes in mechanical equipment and faster interchange of passengers. This would help relieve congestion, he believed, even apart from costly remedies of new construction for subway lines, elevated lines, double-decked streets, and widened streets.[43]

Various remedies for improved passenger accommodation flooded onto the scene beginning in the late 1920s.[44] Electric railway executives had come to understand that their market demanded greater dependability, increased speed, and, mostly because of the invidious contrast of public transportation with private automobiles, increased comfort.[45] In early 1931 the *Electric Railway Journal* featured the Blackhall, a passenger car whose design was initiated and supervised by the general manager of the Chicago, North Shore & Milwaukee Railroad. Designed for a single motorman to operate, it had such improvements as double trucks designed for heavy traffic, a heavy-duty under frame, improved brakes, noise-reduction features, and a pleasing exterior appearance.[46] It was intended "to provide transportation which people will prefer to driving their own automobiles."[47] In Baltimore, the United Railways & Electric Company ordered cars with leather upholstery, ample lighting, and easy entrance and exit. Increasing speed and braking were also engineered into the new purchases.[48] The rate of acquisition was moderate in the early depression years but acquisitions were made nonetheless.[49] The *Electric Railway Journal* also wanted people to see improved cars through the lens of the industry's constantly good service (fig. 5.3).[50]

Several cities initiated plans to better regulate traffic. Detroit's latest efforts were to reduce traffic hazards and economic losses resulting from congestion and to more efficiently use its streets. An eleven-month-long survey submitted in 1927 recommended a uniform traffic signal system that gave appropriate time for pedestrians to cross streets, various regulations on parking such as maximum use of travel lanes, and more efficient use of existing transportation means, including the car, bus, and streetcar.[51] Pedestrian and parking behavior were also studied,[52] and a relocation of streetcars and automotive traffic was proposed (fig. 5.4).[53] Detroit also experimented with a combined express and local transportation service using cars and buses to avoid the high costs in subway or elevated construction.[54] Chicago banned parking in its central business district with the high-minded hope that "this city can continue to grow until it is the greatest in the country, and the most convenient one in which to do business."[55] San Francisco's system for rebuilding the track that the Market Street

FIGURE 5.3. For the Detroit United Railway, the J. G. Brill Company of Philadelphia, Pennsylvania, built ten cars with special safety and comfort features and advertised that they were "successfully combating gasoline propelled competition." They also competed against a traction line in downtown Detroit. "Automobile Competition In Mr. Ford's Own City," *Electric Railway Journal* 71 (June 30, 1928): inside back cover.

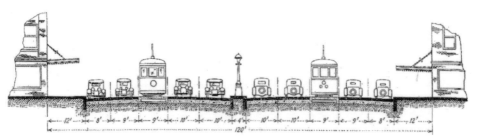

FIGURE 5.4. The 120-foot wide street in Detroit to be rebuilt for easier access between street cars and through traffic without interrupting the roadway's center. "Redistribution of Street Space Proposed at Detroit," *Electric Railway Journal* 75 (July 1931): 362.

FIGURE 5.5. Canal Street, New Orleans, Louisiana, circa 1920. Postcard, authors' collection.

Railway used was featured in a separate article touting the remarkable feat that the work was completed while the street remained open to traffic.[56] The Louisville Railway initiated a system to increase safety on its city's streets.[57] A Houston, Texas, study reported that diagonal parking reduced street access and increased accidents.[58] The *Electric Railway Journal* picked Atlanta, Georgia, over forty-one other applicants for its maintenance award in 1931, which was remarkable in light of the winning Georgia Power Company's goal of lowering operating costs.[59] Cincinnati's concrete loading platforms, whose installation began in 1926, were roundly approved for protecting pedestrians and perceived as superior to the traffic lights automobilists so generally disliked, at least in the early 1930s. "Loading platforms are more than loading platforms," an editorial in the *Cincinnati Post* insisted. "They are isles of safety for pedestrians crossing streets heavy with traffic. . . . If loading platforms were taken away traffic lights would have to be installed at every suburban intersection for the safety of pedestrians. How would the hurrying motorists like that?"[60] Streetcar lines in New Orleans converged on Canal Street, the center of which for some ten blocks was reserved for pedestrians and exiting transit cars (fig. 5.5) But in most cities pedestrians remained at risk as they waited for streetcars at center street. Today passengers still board streetcars there. Street congestion, the *Electric Railway Age* reported in 1931, was exacerbated by drivers' failure to adhere to parking regulations and the frequent practice of having an influential person "fix" the traffic ticket when it was given.[61] Entrepreneurial creativity had not been the problem.

DECLINE

Data in the U.S. Department of Commerce's *Street Railways and Trolley-Bus and Motorbus Operations* published in 1937 discloses the precipitous rise and fall of the electric urban streetcar industry. From 1890 to 1917, the number of companies rose 65.7 percent, the miles of single track operated rose 452 percent, and the number of passengers rose 455.6 percent. Then, between 1922 and 1937, the number of companies decreased 65.7 percent, the miles of track decreased 45.9 percent, and the number of revenue-paying passengers decreased 40.9 percent. Some of the decreased service was taken on by trolley buses (powered by overhead wires or power plants on board).[62]

In 1931 the *Electric Railway Journal* took stock of its subject as "an industry which for more than a decade has had to struggle for bare existence." Automobile competition had been paramount, encouraging "people to move about as never before." Urban life had been "drastically altered." A "new restlessness and roaming spirit" had been stimulated and a "sense of luxury or of personal prestige" pushed the "tremendous increase in automobile passenger-miles and registration." In larger cities the future looked brighter, but in smaller cities "the automobile has proved a greater menace." When income declined, streetcar companies had less capacity to maintain good service and pay for paving construction, track maintenance, and special taxes or operation charges. Streetcars, the journal acknowledged, may not be the best means of transportation, but because smaller cities wish to grow or even survive, certain "comforts, conveniences and necessities" were assumed. Without adequate travel facilities between home and work, schools, and more, small towns could not grow into cities. What businesses expanding the economy would want to move there? Hope dawned for electric street railways as critical contributors to the nation's growth, and the railways' mangers were undertaking improvements that would make their companies' services integral in this ethos of capitalistic and community growth. With the difficulties of the world war over and despite the problems of the Great Depression in play, the future looked bright. "The local transportation industry is in a state of transition" and would in the future be reconstructed "to give a superior service to the communities which depend upon it for their growth, prosperity, and, in fact, for their very existence," the *Electric Railway Journal* closed its testament of faith in September 1931.[63] The journal itself ceased publication at the end of 1931 and in this demise appears in retrospect to have been a bellwether for the electric railroad industry.

Electric trolleys did not become extinct. Ones without tracks contributed in Dayton, Ohio, Seattle, Washington, and San Francisco, California. And in the 1980s interest in electric traction reemerged. It was named "light rail" and appeared most notably in Buffalo, New York, and Portland, Oregon.[64] Nostalgia had driven trolley car preservation among "rail fans" from the 1930s, and after the trolley car virtually ceased operation everywhere, it led in 1967 to the first museum.[65]

JITNEYS

Jitneys embodied capitalism's idealized challenges to established business where improvements for the public's benefit derived from vigorous competition. Jitneys initially had none of the government's limitations assigned to transportation except adherence to local automobile regulation. Individuals with the capacity to purchase and drive these small buses could go into business for themselves and stay as long as they wished. The *Electric Railway Journal* most often mentioned the jitney in negative terms as unfair competition for streetcars (fig. 5.6).

The journal's concern for the troubling new transportation form—only a year old in 1916 on the Pacific Coast—generated a brief report about the region. The jitney quickly became popular, but "as the novelty wears off and the service becomes less and less efficient there is a noticeably increasing demand for more stringent regulatory measures." In Los Angeles, where 350 jitneys operated, the city undertook to have jitneys operated more safely and produce less traffic congestion. Jitney operators considered it a "confiscation of their rights" that they adhere to the regulations in each town through which they pass. Resistance in San Diego to their operation only on certain streets led them to apply for licenses that allowed them to go where they wished. When San Francisco held its exposition, jitneys swarmed into service there and out of Oakland, where the interurban lost $350,000 because of jitneys. In Seattle differences between two factions of jitney operators ended in the demise of both organizations, and Washington's enactment requiring bonds for their operation drove many out of business. Washington also enacted other regulatory provisions.[66] Data revealed that jitneys worked to the trolleys' disadvantage mostly during morning and evening rush hours.[67] In another article, the journal brought to readers' attention that the Spokane, Portland, & Seattle Railroad, a steam line, had been the biggest taxpayer for the nearby automotive highway, but jitneys, dependent on this highway, paid nothing.[68] Four years later the Federal Electric Railways Commission reported that even in cities where trolleys operated at cost, jitneys with

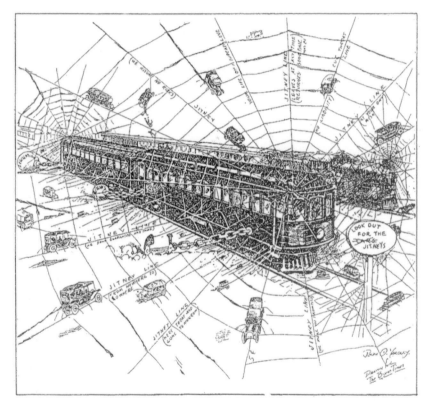

FIGURE 5.6. *The Byron (California) Times* put the jitney's effects graphically and with this caption: "Operators of Railway Lines, Who Have Spent Millions of Dollars, Are Honeycombed by the Irresponsible Jit, Which Networks Their Cars and Tracks, Hampers Transportation and Clamors for Public Support." "Artist's Idea of Jitney Competition," *Electric Railway Journal* 48 (June 24, 1916): 1173.

higher fares competed significantly.[69] Simultaneously, another report to the commission concluded that jitneys began to compete seriously with street railways when the latter came under special taxes and demands for better service.[70]

Jitney owners were prone to wild swings in numbers. In Kansas City, where jitneys first ran in 1915, their numbers spiked to slightly over 240 by spring but plunged during winter a year later to 40. The Street Railway of Kanas City had kept this data about their competition.[71] Such vacillations indicate the considerable degree to which jitneys filled in for other, more regular, or "main-line" transportation forms. Kansas City adopted an ordinance after the war disallowing jitneys to operate on the same streets as trolley cars.[72]

War induced government considerations for electric railways regarding their jitney competitors. The Council of National Defense declared its desire to cooperate with all electric railways and other transportation means for the most efficient, prompt, and economical service. Evidence was uncovered that 5,879 jitneys performed nonessential and duplicate services. The wartime increase in labor witnessed a decline in jitney numbers for several reasons: their operators found higher wages in other businesses or were drafted; operation costs increased; and, coincidentally, increasing regulatory laws around the nation drove them out of business. An assistant manager on the Electric Railway War Board recommended the discontinuance of all jitneys duplicating work. Editors of the *Electric Railway Journal* thought prospects were positive, for the Council of National Defense "and various government agencies [were] beginning to realize the non-essential character of the jitney and to see that, although these [were] declining in number, no time should be wasted in suppressing entirely all useless competition with existing electric railways."[73]

Jitneys performed very ably in other circumstances. An executive of the REO Motor Car Company noted in retrospect that in early 1921, when streetcars faced various challenges, jitneys had rapidly transported workers to munitions plants and shipyards in New Jersey. Their work was literally in defense of life itself. They became important in transporting nearly 200,000 country children to school, and they worked in more than 150 cities and towns nationwide. "Low cost of operation, coupled with the ability to show results under any condition, is making the 'jitney' bus a real competitor against the streetcar," a writer for *Highway Transportation* noted.[74]

The jitneys' sometimes quixotic popularity brought forth ordinances to regulate automobiles hired for service, all apart from "jitney buses" specially constructed for multiple passengers. Connecticut's regulatory law defined a jitney as anything "resembling the service rendered by electric railways."[75] *Electric Railway Journal* all but scolded private citizens who boarded passengers waiting for streetcars. In a case before a New York court in 1924, people were reminded that they were eligible to claim damages for misfortunes that befell them while in the private car.[76] The journal pointed out the case of Davenport, Iowa, where the jitney bus service was declared illegal but the unemployed put their own automobiles into service.[77] In a gentle demeanor the journal reported how the Hot Springs [Arkansas] Street Railway was badly hurt for want of business where, in the words of the headline, "An Autoist Applies the Golden Rule." Improvised jitney drivers—that is, motorists who happened to chance upon people waiting for a streetcar and did them a favor by carrying them to their destinations— badly hurt the street railway business. This was especially vexatious in small-town markets.[78] In Detroit such drivers "infested" streets served by trolley cars,[79] reappearing like the troublesome insects to which they were likened in figure 5.6. The Michigan Supreme Court disallowed jitneys in Detroit in 1926.[80]

Bus Transportation reported in 1932 that St. Louis was rather behind the times among larger cities in its public transportation because it still employed the "service car." Some 350 sedans made up this group, which operated as an independent company, "an outgrowth of former jitney and taxi operations so generally prevalent a decade ago" in the nation.[81] Jitneys operated in Miami beginning in 1919 and as late as 1938. Transit buses provided most of the city's transportation by the end of the 1930s.[82] "Jitney" had become a term of opprobrium for some transportation professionals even earlier.[83] Jitneys were no longer significant enough to win attention when, in 1941, John Anderson Miller published his well-regarded history of urban transportation.[84] However, jitneys had been the short-lived creations of individual owners in the early automobile age; they were driven out of intracity transportation by corporate competitors that backed transportation legislation, first in favor of their own electric trolleys and finally their motor buses.

TRANSIT BUSES

Called the "motor bus" during its introductory years, what is now known simply as the "bus" first gained considerable attention on the pages of the *Electric Railway Journal* just before World War I. But the subject expanded greatly in the 1920s.

Seemingly ever-increasing traffic congestion induced continual reflection on how to profit from intracity transportation needs without simultaneously causing more congestion. An executive of the Mack Truck Company explained that traction lines seemed incapable of handling increased congestion because the lines' capacities were limited to existing rail lines. The motor bus would not only solve the congestion problem but provide a "more flexible satisfactory mode of transportation at a lower cost."[85] More tracks could not be laid without increasing congestion. In bigger cities no room was available for added tracks. Given the fact that streets had limited widths, might the time come when they would be unable to contain the congestion? In small towns the expense of pavement, bridges, grading, and rails would be considerable. Buses were flexible; they could shift to highways. They were faster, enabling them to make more trips in a day, and safer, unloading passengers at the curb. In rural schools, beginning in California, buses were useful and efficient because they easily served schools consolidated over larger, less-populated areas.[86]

Later authors reinforced arguments for the bus' flexibility vis à vis rail-bound streetcars. Buses were better where railway operators wanted to extend existing lines and for service across towns between factory and residential districts, for "belt" lines for passenger convenience and saving time for passengers to urban fringes rather riding a track around urban fringes.[87] A former state official, but an attorney at the time of his published advice, stated that buses were more comfortable (if mounted on pneumatic tires), were safer because they could turn to avoid collisions, and minimized delays in service. "We now have the day of reckoning and it looks as if the rule that a business must run at a profit or be junked would apply to the trolleys," he wrote in *Electric Railway Journal*.[88]

The same writer thought that proponents of trolleys seemed too often to favor their transportation type exclusively and that they would be well advised to employ other types of vehicles to assist trolley systems.[89] Street railways would do well to consider developing motor freight services to areas outside the city, proposed another writer. Buses were viewed as a means of expanding railroad business because they literally afforded "a means of release from confinement to the limits of rail operation."[90] Their use should be coordinated with electric railways, for example, serving as feeders from city outskirts.[91]

Growing favor for the "motor bus" inspired inventive creativity. In 1921 *Highway Transportation* carried an article that in effect publicized the new vehicle's improvements, with solid-tire versions that were as comfortable as those on pneumatic tires.[92] Customers also used them imaginatively; double-deck buses seemed to welcome passengers without automobiles for "pleasure rides" on evenings, Sundays, and holidays, replacing trolleys for this recreation.[93] In retrospect, in 1931, *Electric Railway Journal* saw the 1920s as an era of rapid improvement in bus design: "Interested observers marvel at the rapidity with which changes in bus design have taken place during the past decade."[94]

Articles began to appear showing how various cities used the motor bus. In Los Angeles, an executive of the Pacific Electric Railway looked to California for understanding; he maintained that even in the state where he lived, where trucks benefited from the extensive highway system, trucks should not be permitted to put trolleys out of business.[95] In Akron a corporate company linked to Goodyear Tire and Rubber began service in 1915 between the company's housing development, Goodyear Heights, and factory. *Electric Railway Journal* did not see it as competition for the street railway.[96] Then, three years later, Akron's streetcars became the sole public transportation within the city, and buses ceased operation. There was no talk of later motorization.[97] The aforementioned stalled talks about reforming Detroit's public transportation system resumed in earnest in 1924 and took the form of adopting exclusive use of motorbuses with huge capacity (250 passengers).[98] Back again came streetcars, serving along with buses, according to the *Electric Railway Journal* at the end of 1928.[99] Engineering and passenger amenities on the buses were upgraded.[100]

By the eve of World War II, bus use grew considerably (fig. 5.7). Three executives of trolley services, who were interviewed for *Bus Transportation* in 1928, believed buses and streetcars had respective benefits and should serve jointly.[101] This was upheld by statistics reported in 1928.[102] *Bus Transportation* explained how Durham, North Carolina became an all-bus city in January 1930.[103] The small town of Kent, Ohio, introduced the "taxi-coach" in 1931, a vehicle combining maneuverability in traffic, rapid acceleration and deceleration, passenger comfort, and rapid loading and unloading.[104] The trade literature cast light on several cities for switching exclusively to buses.[105] Most notable was the general shift away from streetcars to transit buses.[106] By 1941, 160 cities out of 411 with populations over 25,000 had electric railroad service; 9 were entirely dependent upon it; and 246 used only motor buses or trolley buses.[107] Private automobiles could not possibly solve the problems of getting to work downtown from the growing suburban fringes, *Transit Journal* admonished its readers on the eve of World War II.[108] Intracity transportation raised the entire subject of urban

FIGURE 5.7. On a postcard of Los Angeles, a city whose landscape automotive vehicles defined more than any other city, a transit bus shows prominently in the lower foreground on Wilshire Boulevard, with tall buildings far off to the side. Author's postcard collection.

development and decline, not only in the thinking of the *Transit Journal* but in the minds of the eleven people who wrote letters to the editor.[109] In the last few months before America's entry into the war, the transportation industry journals focused on the new life-and-death issue of how best to provide public transportation for defense industry workers.[110]

ELECTRIC BUSES

Electric buses powered by overhead wires were welcome to those thinking about future transportation but involved serious impediments rendering them generally impractical for a century or two. *Street Railway Journal* carried reflections on the "trackless trolley" as early as 1904 when an area near Boston introduced one and after one had failed in Scranton.[111] The journal believed they had great potential for rural areas where interurban trains were very costly. Yet, their tires would be expensive, given the quality of the nation's unimproved roads, which were poorly maintained, had steep grades, and were unfit for heavy traffic. They awaited the Good Roads Movement in America.[112] England operated buses successfully for city service, and they fascinated imaginative people in American transportation, one of whom wrote in 1916 about how city authorities affecting streets inspired promise.[113]

Beginning in 1921, the transportation trade journals fixated on the electric bus. A general treatment of all bus transportation forms reported on the electric variety's costs as between those of the motor and rail variety, its overhead wires and vehicle costing more than the motor bus but with lower operating costs.[114] New York's experience with the smoked-filled tunnels caused by steam trains pointed to the advantages of electric power.[115] When in the spring of 1921 the Virginia Railway & Power Company put a "trollibus" into operation on the streets of Richmond, Virginia, the *Electric Railway Journal* declared it to be the first in use, aside from early experiments.[116] Schenectady tested a new trolley bus in the fall of that year.[117]

The Staten Island Midland Railway introduced trolley buses in 1921. These provided transportation to the previously "inaccessible" Linoleumville and Staten Island's premier Sea View Hospital. Construction costs were considerable: $37,000 to $60,000 per mile for single-track construction versus $4,000 for trackless construction. Additionally, trolley buses did not grind down the roadways; their wear and tear was like that of automobiles.[118] New York City initiated feeder trolley bus service from those places into the city.[119] Cohoes, New York, varied the system when it adopted trolley buses to replace electric trolley cars that had taken passengers from the city into a belt line around it. Track and paving expenses were removed, while fares remained the same.[120] Manufacturers with successful records in mass production took notice. The Packard Motor Car Company and the Westinghouse Company saw advantages in these new devices: maneuverability on city streets to avoid accidents and an end to track noises. Gasoline bus design, not streetcar design, ruled the companies' plans.[121]

"Trackless trolley bus will undoubtedly be a factor in the transportation of the future." With these words the *Electric Railway Journal* declared with certainty at a time of rapidly evolving transportation modes, both disappearing and rising, that the advantages of the trolley bus were undeniable: practical for feeders to electrical lines, practical for towns that outgrew electrical lines, practical for towns that grew in unanticipated directions. The trolley bus was comparatively inexpensive and practical.[122] A decade later, however, readers of the *Electric Railway Journal* were admonished not to think of the trolley bus as the panacea.[123]

Trolley buses served to variable extents where they were employed. In 1928 Salt Lake City, Utah, announced a plan including them in a transportation system operating buses and trolley cars because, while the latter were under consideration to be dropped from service, trolley buses were welcomed as a quiet, high-speed, comfortable, and odorless alternative with comparatively inexpensive operating costs.[124] Subsequent analysis, however, revealed that the amount of travel the trolley bus would offer did not justify the use of a vehicle so large, not to mention the cost of installing and maintaining the overhead cable system. So, in 1935 Salt Lake City instead added more transit bus service.[125] Here was an example of a desired result being reported as fait accompli, when, in fact, Salt Lake City retained its trolley buses; this the general manager of Utah Light & Traction Company sternly assured *Bus Transportation.*[126] *Electric Railway Journal* reported that the numbers of trolley buses in use rose after 1921, declined from the mid- to late 1920s but rose sharply in 1930.[127] But was the trolley bus experiencing a huge swing in its favor? Only 167 were at work nationwide in 1930.[128]

FIGURE 5.8. A postcard, postmarked 1943, proudly shows a trackless trolley in Seattle, where the mode, all at once, had taken over the intracity transportation. Postcard, authors' collection.

During the following decade, trolley buses did experience an incredible growth in use. At the end of 1940, 2,836 were in use, approximately a 28 percent growth over the previous year.[129] *Transit Journal's* editor anticipated growth of those in use to about 3,150 by the end of 1941 and far more over the next decade. Seattle changed over to trolley buses for its only form of public transportation during 1941 (fig. 5.8).[130]

Convenience, flexibility, cost, and appearance wedged the bus's automotive advantages, in the form of the trolley bus, onto some streets in some cities. Surpassing trolley cars, trolley buses, which blended the modes of electrical mass transit and gasoline-powered mass transit, emerged as an interesting element in intracity transit at the middle of the twentieth century. They symbolized some of the old and some of the new.

Through the first forty years of the twentieth century, urban and nearby suburban expansion had winnowed through several forms of intracity transportation. Advantages—maneuverability, relative noiselessness, cleanliness, and looks—had put trolleys behind electric streetcars, transit buses, and electric buses. Technological advances leading to the inherent superiority of one mode of transportation over another were not the only factor in play. The jitney was an automotive vehicle, but its owner and operator drove for a reasonable profit without harboring hopes of dominating the market as did the owners and managers of the other transportation forms under corporate auspices. Jitneys were swept off the streets. Traffic congestion, an unforeseen and undesirable byproduct of a successful and growing city, was a problem that jitneys worsened but certainly did not cause by themselves. Intracity transportation raised important questions and were themselves important issues in restless city life.

Chapter 6

INTERURBAN RAILWAYS

E lectricity was first applied to transport in the United States for powering city streetcars beginning with Richmond, Virginia, in 1888. A decade later, the new technology was also being applied to travel between cities but more generally from cities to outlying towns and rural areas. The industry created was viewed as complementing the nation's steam railroads but, nonetheless, as one also ready to compete with them aggressively. A decade on, the steam railroads themselves began to electrify selected routes, especially in the heavily traveled rail corridors of the Northeast. Electrified railroad lines continue in operation to this day, although much diminished in their total mileage, whereas the nation's interurbans (or "traction lines" as they were frequently called) were abandoned almost as quickly as they evolved. The years 1900–1907 witnessed a virtual explosion in interurban mileage. Most interurban railways were lightly built and, accordingly, used lightweight equipment. Tracks and rolling stock, therefore, tended to deteriorate quickly. They operated more like streetcar lines than steam railroads in that they sought mainly passenger traffic, with what little freight they carried tending to be insufficient to subsidize passenger service, as was the case with the steam railroads. After 1916 the rapidly increasing popularity of motoring put interurban ridership in rapid decline. With revenues falling, interurban companies could not raise the capital necessary to repair tracks, replace cars, and otherwise improve passenger service, thus to combat automobility. Competition from buses (and from trucks for those companies that did handle freight) was partly to blame, but it was mainly the popularity of the private automobile that undermined the interurbans.

We begin with three assertions. First, had Henry Ford developed his Model T only a few years earlier (that is, had mass automobility come to the United States a bit sooner), the interurban industry might never have happened. What Americans were looking for was speedier and more convenient transportation between cities and the outlying hinterlands. Electric transit cars running down the centers of or immediately along rural highways, as most of the early lines did, offered that. In comparison to what passenger trains provided, interurban cars ran more frequently, and usually stopped on demand wherever and whenever passengers requested. Automobiles, of course, were completely at the motorist's command and ordinarily traveled at even faster speeds in open country. Automobiles could go anywhere and everywhere. And they did. Additionally, automobile ownership quickly became a badge of social advancement, a matter of social mobility and not just geographical mobility. Mass automobility proved the interurban's undoing. That is the story we emphasize herein. But had inexpensive and dependable motor vehicles come sooner, there likely would not have been a traction industry to undo.

Second, the coming of an interurban line represented for most localities in the United States the initial step toward improving rural roads. In a sense, the interurban industry led the Good Roads Movement, perhaps more so than did the bicyclists. The nation's "wheelmen" merely lobbied for highway improvements. The interurban companies actually brought physical change to roads when they placed their tracks in or alongside them. The intention, of course, was never to run private vehicles. Quickly discarded was the "rail highway" idea whereby anyone might operate a vehicle on roads improved with rails. The interurbans were built with commercial intent—that is, profit-making through private enterprise. In many localities such intent, indeed, proved necessary before local or state governments could be convinced to invest in improved roads.

Third, the steam railroad industry did not welcome the interurban and indeed sought to block the building of interurban lines in the courts—for example, through litigation to prevent interurban tracks from crossing their rights-of-way. Railroad executives might not have immediately recognized the automobile's competitive implications, but they clearly understood that the new traction lines would divert passenger traffic. Interurban apologists argued that the steam railroads were poorly equipped to provide inexpensive and frequent short-distance passenger travel in and out of cities. That was the niche that the new industry intended to fill. In so doing, the new electric lines would serve small towns and villages in localities where the railroads either did not provide service or only poor service. Of course, many interurban lines were actually built closely paralleling steam railroad rights-of way. The railroads were correct in discouraging that new threat. If they could not stop interurban line construction, then they could, as most did, refuse to exchange passengers and freight through coordinated schedules and shared terminals. Indeed, at first there was very little cooperation.

That interurban construction might hasten highway development occurred early to Good Roads promoters. H. W. Perry, writing in the *Good Roads Magazine* in 1901, noted that some seven thousand odd miles of interurban track, equal to the mileage of the steam railroads, was underway in Ohio, some 240 interurban companies having been organized in the previous three years. Traction lines, he noted, "must be studied for the purpose of determining the effect they will eventually have upon highway travel, whether toward increasing or decreasing its volume, or as reducing the length of haul by horses and wagon, and changing its direction." Interurbans, he added, transported "passengers from town to town, or between points on the road, as comfortably, more cheaply, and almost as quickly as the steam cars between stations, and far more quickly and agreeable than the trip, long or short, can be made by horse and buggy and wagon." (In this regard, one remembers that the Good Roads Movement well predated mass automobility.) Besides, Perry added, passengers in interurban cars could carry "trunks, milk cans, farm produce, farm implements, and a miscellaneous lot of small merchandise." Some lines also carried "Uncle Sam's mail bags." Thus Perry surmised that "improving steam- and gasoline-powered autos, buses, and trucks might prove superfluous."[1] Of course, he was wrong.

THE INTERURBAN'S RAPID RISE

In 1889 there were but seven miles of track that could be said to be "interurban."[2] An interurban railroad differed from streetcar lines in many important respects, although the difference between them was sometimes quite fuzzy. Indeed, many of the earliest lines were little more than rural extensions of city streetcar opera-tions, especially in New England, where, for example, Boston's streetcars trundled well beyond the city limits into the suburbs. However, a proper interurban line was one that mainly traversed rural countryside in addition to city space and was, therefore, much longer in length, requiring faster running. Thus they used heavier equipment on slightly heavier tracks. In particular, interurban cars moved faster when built on rights-of-way separate from highways, which became in-creasingly common as the industry matured. In 1900 there were some 560 miles of track laid, but from 1901 through 1904 an average of some 1,300 miles were laid each year. True, the amount constructed in 1905 was cut nearly in half by the financial panic of the previous year, but between 1906 and 1908 development continued apace, averaging some 1,200 miles annually. Then the Panic of 1907 wrought a sharp drop. Between 1909 and 1920 (when for the first time no new mileage was added), the annual average for new trackage laid nationwide stood at just 427 miles. Only 158 additional miles were built in later years.

The impact wrought by improved automobiles, buses, and trucks, along with the improved highways to serve them, was fully evident. Capital for expansion became all but impossible to obtain and, for that matter, financing to improve existing rights-of-way, modernize rolling stock, and improve power and termi-nal facilities was exceedingly difficult to come by. Trackage began to be aban-doned and then entire systems closed down. Track abandonment averaged some 46 miles per year between 1908 and 1920. The early enthusiasm for interurban construction had produced superfluous lines requiring such adjustment. But between 1921 and 1929 an average of some 456 miles annually were abandoned. During the same period, an average of only 23 new miles were built per year. The stock market crash of 1929 and the Great Depression of the 1930s that fol-lowed placed most of the nation's interurban companies in receivership. Some were merged into larger systems, but most were liquidated. Between 1930 and 1940, some 749 miles were abandoned each year and no new track added.[3]

Traction companies, as with other business concerns, were financed through stock and bond issues. They were financed, that is, largely through debt. The size of a company's capitalization reflected both anticipated costs of construc-tion and operation, and anticipated revenues based on expected traffic. But the great enthusiasm for electric railways led developers to underestimate costs and greatly overestimate anticipated traffic. The latter was substantially caused by failure (or unwillingness) to recognize the automobile's potential popularity.

As a consequence, the industry was never really fully profitable. Few companies ever paid a stock dividend. Economists George W. Hilton and John F. Due, whose book *The Electric Railways in America* remains the classic history of the industry, established that at least 10 percent of the interurban mileage at any one time was owned by companies in receivership. Receivership eventually led to foreclosure and reorganization, with the control of a company typically passing from stockholders to bondholders. As Hilton and Due pointed out, interurban ownership fell into several patterns. About a third of companies were completely independent, having been established by an investment syndicate unrelated to any other enterprise. Some were subsidiaries of electric power utilities. Some were controlled by city transit systems, and a very few were actually controlled by a steam railroad.[4]

Among the independents, for example, was the large Union Traction System of Indiana. Among the power company affiliates was Illinois Traction (eventually Illinois Terminal Railroad). The Detroit United system, on the other hand, was controlled by what became the Detroit Street Railways. Holding companies controlled many lines, power companies often serving that function—for example, Samuel Insull's Commonwealth Edison, which came to control the Chicago, North Shore & Milwaukee, the Chicago, South Shore & South Bend, and the Chicago, Aurora & Elgin. Many interurban companies changed ownership several times. Some, like the Detroit, Toledo & Ironton Railroad, even ended up in the hands of auto interests. So it was that Clement Studebaker of South Bend, Indiana, bought from William B. McKinley of Champaign, Illinois, the Illinois Traction system. McKinley then retired from business to successfully run for the U.S. Senate. The Cities Service Company, one of America's largest gasoline retailers, took control of interurbans in Ohio, Indiana, Oklahoma, Colorado, and Washington.

The largest number of interurban projects were undertaken in Ohio. Indeed, about 54 percent of the nation's total traction mileage was eventually built in that state. As Hilton and Due noted, ninety-two traction companies were incorporated in Ohio in 1901, forty-seven in 1902, and forty-six in 1903. Had all the proposed lines ultimately been constructed, Ohio would have had some 9,000 miles of interurban railroad. As it was, Ohio did end up with some 2,800 miles of line.[5] Indiana (with 1,800 miles) was second in mileage, Pennsylvania (1,500 miles) third, New York (1,100 miles) fourth, and Michigan (981 miles) fifth.[6] New England had considerable mileage, but since most of it was owned and operated by city streetcar lines, it was difficult to differentiate it from streetcar mileage. New England traction systems, as opposed to those in the Midwest, operated heavily in areas of high population density, and thus were very streetcar-like in terms of rights-of-way, rolling stock, and operations.

Interurban lines were designed mainly for single-car passenger trains. Locals involving frequent stops were the order of the day. Interurban express trains were few, requiring as they did numerous passing tracks if not extensive stretches of double track. But some lines were built expressly for them—for example, the Lake Shore Electric Line, which was constructed to enable a four-hour and twenty-minute express-train ride between Cleveland and Toledo. Suburban service around Chicago also evolved with express as well as local service. But most tractions' lines depended mainly on farmers and their families traveling to town to shop or, perhaps, sell produce, and, of course, on salesmen and others traveling on business. Frequent service was of the essence, not just several trains a day but one every hour. Indeed, traction franchises issued by local governments often necessitated such frequency. Franchises, especially those that allowed street-running, required interurban cars to make frequent stops—for example, at all major road intersections. This is not to say that marketing long-distance travel was never in a company's plans. As early as 1904, several Ohio traction lines (the Lake Shore Electric, the Toledo & Western, and the Detroit, Monroe & Toledo) began coordinating schedules and issuing through tickets for travelers between Cleveland and Detroit, a distance of some two hundred miles. "The arrangement between the companies is a simple one," explained *Street Railway Journal*. "Each road supplies the others with its tariff sheets and time tables, and agrees to accept the tickets issued by the other companies. The round-trip ticket issued is made up of various coupons, and is precisely similar in general appearance to the coupon strip tickets used on steam roads."[7]

Even so, most interurban customers rode for only short distances. Most confined their travel to but one line per trip. Indeed, the main rationale behind traction-line development was providing locals with cheaper and more frequent train service than the steam railroads did, and/or providing service where the steam railroads did not. A case in point was Ohio's Dayton, Covington & Piqua Traction Company. Thirty-four miles long, the line was initially intended to connect Dayton and Covington to the north in the Stillwater Valley (fig. 6.1). But a six-mile connection was then built between Covington and Piqua. As the *Street Railway Journal* reported, "The valley is lined with small farms given largely to market gardening, and the produce is old in Dayton and Cincinnati. Before the building of the electric railway the people were obliged to depend upon a branch of the Cincinnati, Hamilton & Dayton [steam] Railroad, which ran but two trains a day. "The steam road takes a circuitous route into Dayton, and although the electric line makes no pretense at being a high-speed road, its cars give better time into the city."[8] Then an additional line was contemplated to Versailles, Ohio, even farther to the north. Such was the planning that brought most interurbans into existence, with the success of an initial line prompting a series of incremental line expansions.

The mania for interurban building centered largely in the Middle West (fig. 6.2). In Ohio, multiple traction lines radiated out of Cleveland, Columbus, Dayton, Lima, and Toledo. In Indiana, Indianapolis was the focus of twelve lines

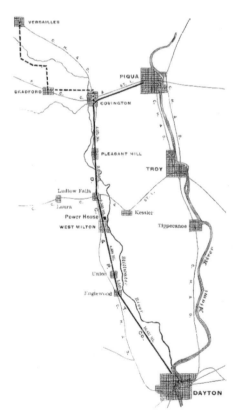

FIGURE 6.1. The Dayton to Piqua, Ohio, Traction Line. "The Dayton, Covington & Piqua Traction Company," *Street Railway Journal* 24 (Sept. 17, 1904): 390.

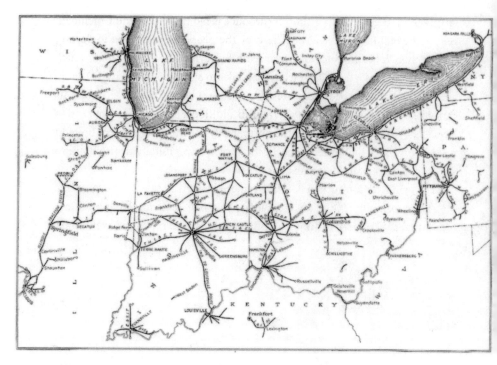

FIGURE 6.2. Interurban Railroads of Midwest, circa 1915. A. B. Cole, "Electric Railways Are in a Position to Haul More Freight," *Electric Railway Journal* 51 (May 11, 1918): 916.

connecting to Fort Wayne, Louisville, Richmond, South Bend, and Terre Haute, among other cities. In Michigan, Detroit was the principal hub. In Illinois, lines radiated out of Chicago. The McKinley Lines downstate connected Danville and Champaign-Urbana on the east, as well as Peoria at the center of the state, with St. Louis. Lines in Ohio interfaced with those of Michigan and Indiana and thus could exchange traffic through Toledo, Lima, and Dayton, respectively. But Chicago remained isolated from the others save for the nearby city of South Bend. Downstate Illinois's extensive mileage was never linked with that of the Hoosier State. Lack of linkage clearly reflected the local character of most interurban projects, although eventually many lines were, indeed, consolidated into larger systems. The Census Bureau prior to 1912 did not tabulate mileage figures for interurban traction lines. In that year, however, it did report some forty-one thousand miles of electric railroad in operation in the United States, sixteen thousand miles of which were classified as interurban. Of that mileage nearly one-half lay in Ohio, Indiana, Illinois, Michigan, and Wisconsin.[9]

TRACKS AND RIGHTS-OF-WAY

In Indiana track weights increased from 20 or 30 pounds per yard to 90 and 100 pounds, the heaviest in the state being laid on the Chicago, South Shore & South Bend at 112 pounds. Save for a very few miles, all of Indiana's interurban trackage was of standard gauge (four feet, eight and a half inches).[10] Rights of way varied. Of course, the least expensive place to lay interurban track was in or along established streets and rural highways. That is why so many early projects did so. The problem was that tracks were forced to follow the gradients and curvatures of the established roads, often making for slow train operation. As roadways were often poorly drained, ties upon which the rails rested tended to deteriorate quickly, necessitating frequent repair that the companies became increasingly hard pressed to finance. Of course, interurban cars on such rights-of-way had to contend with ever-growing amounts of motor traffic. Early on, most interurban lines reached city downtowns on streetcar tracks, effecting "street-running."

Pictured are two sections of the Terre Haute, Indianapolis & Eastern Traction Company's line between Terre Haute and Indianapolis (figs. 6.3 and 6.4). The first, taken around 1920 at Lewisburg, Indiana, shows light rail installed down the center of the National Road, what became U.S. 40. Note the automobile and commercial garage to the right. The second, taken near Indianapolis a few years later, shows an interurban car operating on a fully independent right-of-way, one separated by viaduct from an overhead crossroad. In towns and cities where streetcar tracks were not available for street-running, tracks had to be laid. Then traction companies were usually responsible for paving streets not only between the rails but several feet on either side. Usually the light rail laid had little to recommend it except low cost.[11] Under the weight of heavy traction cars, light rails and cross ties quickly settled in, with the tracks becoming rough and irregular. Through the early 1920s, most new traction lines involved totally independent rights-of-way. And there construction tended to be of higher quality. Many traction rides, however, involved numerous kinds of right-of-way. The Boston & Worcester Street Railway, popularly dubbed "The Trolley Air Line," opened for traffic in 1903. A trip from Boston actually began on the Boston Elevated Railway Company's surface line at Chestnut Hill. Then it continued west

FIGURE 6.3. Interurban line at Lewisburg, Indiana, circa 1920. Postcard, authors' collection.

on track laid on the Shrewsbury Turnpike to the bank of Lake Quinsigamond just east of the Worcester. It ended on the Worcester Consolidated Street Railway Company over city streets to the downtown of that city.

Traction companies found themselves facing ever-inflating track-repair costs and at the very time when revenues began to free-fall as the automobiles made greater and greater inroads on passenger traffic. And often the traction companies could not repair their tracks even when funds were available.

Where interurban cars ran on streetcar tracks, traction companies paid a fee and were fully dependent on streetcar companies to use those payments to maintain rights-of-way. Many did not. "The city railway occupies public streets under a franchise, limited in term, and with obligations relating often to track, paving, sprinkling, speed, service, and fares," wrote Theodore Stebbins in the *Proceedings of the American Street Railway and Interurban Railway Association* in 1906. The limited-term franchise defeated its intended purpose, he noted. "It repels capital seeking a permanent investment at low interest rate. It encourages speculative management at the expense of physical property and its service."[12]

In 1909 the average revenue per passenger on Ohio interurbans was only 8.3 cents, a figure largely maintained through World War I.[13] Thus the low average receipts per passenger were reflected in earnings per car mile so small that,

FIGURE 6.4. Interurban car near Indianapolis, circa 1920. Postcard, authors' collection.

right from the beginning, they boded ill for the future. In 1909 revenue per car mile for Ohio's interurbans was 25.8 cents. Since operating expenses averaged 15.1 cents, operating ratios might have seemed favorable, but the rapid rise of maintenance costs relative to receipts caused this ratio to move steadily against the traction companies. By 1920 average revenue per car mile in Ohio had risen to 45.7 cents, but at the same time operating costs had reached 36.1 cents.[14] Costs were that low, in part, because maintenance was being postponed. Companies cut train speed rather than repair track. With revenues low, traction companies found it difficult not only to renew track, but also to modernize rolling stock. Slowing travel times in increasingly uncomfortable interurban cars only discouraged ridership, making motor-vehicle use an ever more attractive alternative.

POWER

The development of high-voltage power transmission (not only direct current but especially alternating current), along with the development of rotary converters and improved transformers, made interurban service possible. So rapid was the improvement that it, in itself, greatly stimulated the early rush of capital into the booming interurban industry. Investors did not want to be left behind. As one observer objected, "Funds for the projection of interurbans obtained through directing attention to the development of the art rather than to the development of traffic or the difficulties thereof."[15] Most traction companies supplied their own electricity. Self-supply came as a matter of course for traction companies that were subsidiaries of public utilities. Many traction companies, on the other hand, supplied surplus power to municipal and private customers. The marketing of direct-current power, however, was vastly curtailed when the spread of home appliances created demand for 60-cycle alternating current. Transmitting DC power required the use of storage batteries in substations along a traction line. The batteries were connected to trolley wires through motor-generator sets called boosters. These systems proved complex and costly, and by 1910 most had been superseded by alternating-current transmission.[16] AC generating plants produced three-phase alternating current capable of serving some two hundred miles of track. Three high-tension wires were usually carried atop power poles, the cars taking power off a single wire suspended below. Relatively few traction companies distributed power from third rails. Alternating current transmission was not without its problems; for example, "line drops" or transmission losses were greater than with direct current. Some traction companies, therefore, remained with direct current.

Most interurban power houses were modest-sized structures very much like the municipal power-generating plants then common to small towns. Subsidiaries of large public utilities, of course, drew their power from mammoth central power stations. Power-boosting stations were small, single-story structures of reinforced concrete, brick, or frame, but frequently they were integrated into passenger stations, freight stations, or even interlocking towers. Some substations were portable, allowing them to be moved according to changing traffic demands. Retired interurban cars were often used for the purpose.

In 1916 the Union Traction Company of Indiana furnished energy for light and power to twenty-six municipalities varying in size from two hundred to fifteen hundred people. The Illinois Traction Company allowed grain elevators to draw electricity directly off its interurban transmission lines. The Fort Wayne & Northern Indiana Traction Company operated a power subsidiary, the Wabash Valley Utilities Company, which distributed electricity to communities located both along and off its tracks.[17] Wholesaling power required operating interurban power houses throughout the day rather than for the fourteen to eighteen hours during which trains actually ran. Selling surplus power did help balance demand, with power plants no longer required to cope with the unevenness of peak-power requirements so problematical with public utilities. Aside from the revenue it provided, contracts for power tended to improve public relations in the towns up and down an interurban line. And it helped keep railroad employees busy. "It appears to be customary for the electrical engineer and his assistants to devote a portion of their time to supervising the lighting and power development, while the substation inspectors, ticket agents and attendants read the meters and collect the bills," according to *Electric Railway Journal*.[18]

INTERURBAN CARS

Interurban cars varied from small local (or short-run) cars to those of the "limited" (or express) trains used on long-distance runs. The former dominated the field early and were usually of wood-frame construction on steel platforms, the typical car being about thirty-five to forty feet in length and seating thirty-two to fifty people. Maximum speeds were twenty-five to thirty-five miles per hour. The latter came later and were largely of all-steel construction, save for interior finish. In 1904 cars of the Indiana Northern Traction Company, which linked the city of Wabash with Marion, were designed to be operated always in the same direction (fig. 6.5). At the front end was a "motorman's cab" separated from seats behind by a partition. At the rear was a smoking compartment closed off from the remainder of the car. Located there was a space heater that warmed the entire car in cold weather. One entered through doors placed front and back on the car's platform side.[19] Many companies, however, preferred cars where a motorman could operate at either end, thus eliminating car turning. Some companies operated combination passenger and baggage cars. Ordinarily, cars operated singly, but where traffic warranted they were sometimes coupled together. Also common were dummy trailers pulled by powered units. Some interurban lines ran open flat-bed cars in the summer months.

On long distance trains that ran at night—for example, those of the Interstate Public Service line between Indianapolis and Louisville—parlor or chair cars (with soft cushioned seats rather than benches) and even sleeping cars (with Pullman-like berths) were provided. Sleeping cars, however, were usually trailers, the thought being that they would be quieter without motors. As described

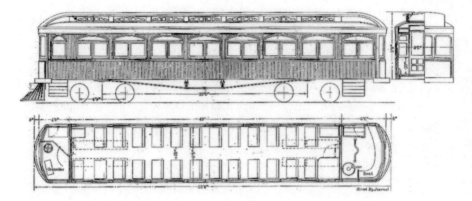

FIGURE 6.5. Interurban car, Indiana Northern Traction Company, 1904. "The Indiana Northern Traction Company," *Street Railway Journal* 24 (Sept. 10, 1904): 365.

in the *Electric Railway Journal,* the Interstate's sleeping cars were divided into ten sections with berths capable of accommodating up to forty passengers. Six feet and four inches long and three feet wide, berths were larger than those on steam railroads. "The ladies' dressing room at the front end of the car, with a passageway to one side," the trade journal noted, "is somewhat larger than that found on Pullman cars. A three-section mirror is located on each of the two side walls. Below each mirror is a dressing shelf and a hair receiver. Three low-back, upholstered chairs have been provided. A corner washbowl is directly under one of the windows and a dental lavatory is between the two side windows. The finish of the woodwork and walls below the belt rail is light mahogany."[20] Baggage compartments in passenger cars gave way to baggage cars, especially when express-mail and/or package-freight service were emphasized. Motorized freight cars (called "box motors") also came to the fore as some interurban companies, faced with declining passenger traffic, turned to carrying freight to supplement revenue.

TERMINALS

With street-running in small towns, most interurban cars merely stopped mid-street in the main business section, with tickets being sold by one of more local retailers. A storefront along Main Street might perhaps be set up as a station with ticket counter and waiting room. Where interurban cars operated on a right-of-way separate from a street, a small station or perhaps only a loading platform might be provided, with tickets sold on the interurban car itself. In cities, on the other hand, depot buildings were usually provided, either owned and operated by a specific company or shared with other companies as a "union terminal." The traction terminal in Indianapolis was the largest built specifically for interurban use (figs. 6.6 and 6.7). In an article filled with statistics, the *Street Railway Journal* noted that "the building is nine stories high, fronts 170 ft. on Illinois street and 70 ft. on Market Street, and contains 250 office rooms, six store rooms and a spacious basement finished in marble." The ground floor was devoted to retail spaces, a café, a baggage and check room, ticket office, and waiting rooms, one for women and children. The train shed was "190 feet long and 175 feet wide, containing nine tracks." The structural ironwork was "both massive and artistic." The floors were cement, and the arrangements for the convenience of the passengers were "as complete and as safe as possible." The terminal was located a block from the state capitol.[21]

Bernard Meiklejohn, writing in the *World's Work* in 1903, also celebrated the Indianapolis Terminal, although it was not, he said, "so complete a structure as a great railroad station, for the cars run in at one end and out at the other." Then he described how it was to travel by interurban:

FIGURE 6.6. Traction terminal, Indianapolis, circa 1915. Postcard, authors' collection.

FIGURE 6.7. Traction terminal, Indianapolis, circa 1920. Postcard, authors' collection.

You step into a long commodious local car, provided with smoking and toilet apartments, and the car is off. The porter leads you to a comfortable wicker chair which you have engaged beforehand, and asks you if you will have lunch. The waiter serves you from a well-stocked kitchen such a meal as you might have in a Pullman buffet car, and when you have lighted your cigar in a cozy leather chair in the carpeted smoking room you are bowling away for a run of 108 miles [to Dayton] that will take but a little more than four hours. . . . You speed across country in a comfortable apartment with broad plate-glass windows and mahogany woodwork—for the car closely resembles a railroad coach—and as you look out across the flying fields your sensation is hardly different from that of riding a train, except that you smell not smoke and no cinders make your blink.[22]

The freight house located behind the Indianapolis passenger terminal was doubled in size in 1911, and then enlarged again in 1924, this time to handle trucks as opposed to horse-drawn freight wagons.

TRACTION PARKS

Traction companies established picnic grounds, beach resorts, and amusement parks as a means of increasing passenger traffic. One of the largest was Norumbega Park at Auburndale (now part of Newton, Massachusetts) some ten miles west of downtown Boston. It was serviced by the Middlesex & Boston Street Railway Company, which extended out along Boston's Commonwealth Avenue well into what was then open country. It was another one of New England's many combined streetcar/interurban operations. The park featured an outdoor theater, a zoo, and boating and canoeing on the Charles River. Opened in 1897, an estimated 6 million customers patronized the park in its first seventeen years. An admiring trade-journal reporter wrote: "The more boisterous amusements of the beach resorts on the Boston North and South Shores have not been found necessary here, and the reputation of Norumbega Park rests upon its satisfaction of more esthetic demands for entertainment than are associated with resorts catering to rough-and-tumble pleasures." In 1914 a round-trip ticket to the park cost twenty-five cents.[23] Some traction lines were almost totally dependent upon resort traffic. In upstate New York, the Fonda, Johnstown & Gloversville Railroad owned Scandaga Park in the foothills of the Adirondacks northwest of Schenectady. "The tastes and demands of the long term summer resorters, who come to us for periods varying from two weeks to a season, call for the concerts which are given three times a day by a first-class eight-piece orchestra stationed on the veranda of our large hotel," reported the railroad's general passenger agent. "We also maintain a standard nine-hole 3090-yd. golf course which has grass tees and regulation circular greens. And there were tennis courts also."[24] One might theorize, however, that it was, in fact, the resorts with the "rough-and-tumble" entertainments, rather than those with politely elite ones, that served interurban interests best.

DECLINE

Interurban fortunes began a slow but steady decline after 1910. As short lines faltered, there was some recourse through their consolidation into larger, healthier companies, which then used them as feeders to their own operations. Mergers that involved recapitalization often involved name changes. One was the Ohio Electric Railway formed in 1907 through merger, purchase, and leasing some seventeen struggling traction companies, forming a system over six hundred miles in extent. In 1921, reduced to some 460 miles in length, it, too, entered receivership. The reasons given were as follows: (1) inability to raise rates on passenger tickets, given municipal franchise mandates, (2) inability to overcome serious damages to track caused by flooding seven years earlier, (3) competition wrought by private auto ownership, (4) the rising cost of ordinary track renewal, especially on urban streets, and (5) substantially increased labor costs. Mainly it was the company's failure to obtain rate increases and thus sustain revenues commensurate with increased costs of operation.[25] According to George Hilton and John Due, the year 1911 was the first one in which interurban officials began to emphasize their unsatisfactory revenue stream. According to Hilton and Due, "It was attributed to two major factors: first, to the fact that many lines had been built in anticipation of traffic increases that had not materialized; and, second, to the fact that fares were set too low in relation to costs, which were tending to rise. Commonly, the promoters of the roads seem to have overestimated revenues and understated costs, perhaps in part deliberately in order to raise capital."[26] From the beginning, interurban fares had been set from 1.5 to 2 cents a mile. Steam-railroad fares were between 2 and 3 cents.

But it was the automobile's coming that had had the decisive impact. It was in 1913, noted Hilton and Due, that there began to be "some perceptible effect" in the volume of business loss to motoring, especially in suburban service. Sales of automobiles, they pointed out, rose from some 63,000 in 1908 to 356,000 in 1912, and to 900,000 in 1915.[27] Nonetheless, so long as cars remained unreliable and rural highways remained primitive, interurban managers could not and would not take the automobile as a definitive threat. Indeed, for years many continued to assume that motoring would prove to be nothing more than a passing fad. Such was not to be. Instead, the interurban proved to be the fad. The life expectancy of the average interurban line proved to be a mere twenty-eight years. The average ranged from a high of thirty-three years in the Midwest to a low of twenty-four years in New England.[28] A 1924 article in *Electric Railway Journal* summed it up:

> When we recall that a large proportion of the interurban mileage in this country was built before an opportunity had been afforded accurately to estimate the amount of business which could be commanded; when we remember that a large proportion of the interurban mileage was unable to earn a fair return on the capital invested even in pre-war and pre-automobile days; when we take into account the tremendous increase in the cost of labor and materials which has occurred in the last decade and give consideration to the reduction of business which the privately owned automobile and the motorbus have caused, we can readily understand why the interurban railways have come to such grave pass.[29]

MOTOR BUSES

If the motor bus could be cited as an interurban competitor, so also might it be, when operated by an interurban company, a means of financial redemption. So also could the motor truck offer redemption for interurban companies seeking added freight business. Operating motor buses provided interurban companies with important advantages. Interurban service could be extended without the expense of added right-of-way. With financial conditions worsening, unprofitable lines, especially branches, could be eliminated and replaced with buses operating over public roads. Indeed, some companies facing liquidation, particularly those that ran on urban streets and/or rural highways, opted to shut down electric service, replacing interurban cars completely with buses. It was very much the same story that played out with city streetcars. It was quickly realized, however, that buses in open country were not as effective as buses in cities, traffic demand being so low. Motorists in small towns avoided the new buses for the same reasons they had come to avoid the interurbans. Compared to the use of private autos, the aging interurban cars simply lacked convenience, speed, mobility, and comfort—personalized mobility, in other words. There was no pride of ownership. Buses were little better.

At first, buses were little more than enlarged automobiles. Indeed, many were used automobiles refitted in various ways to carry additional passengers. Then came bus bodies placed on truck chasses. But by the mid-1920s, buses had become quite sophisticated as we discuss in chapter 8. In size and function, they were vehicles fully divorced from their auto predecessors. Commercial bus companies initially sought to operate routes that closely paralleled interurban lines since that was where local travel patterns were well established. But many states moved to protect the interurbans by refusing bus franchises for such routes. The Ohio Public Utilities Commission, it was reported in *Bus Transportation* in 1936, "has established precedent for refusing permits to bus companies for service which competes with adequate rail service. Commission denied requests of Buckeye Stages, Inc., and Central Greyhound Lines, Inc., for permission to establish local bus service between Toledo and Cleveland."[30] Not trusting to governmental policy, many interurban companies started their own subsidiaries to run buses on parallel routes if only as a means of self-protection. Buses had come to stay. Thus the *Electric Railway Journal,* the leading trade magazine covering the interurban industry, inaugurated in 1921 a special monthly supplement focused on buses.[31] In 1921 twenty-five electric railway companies were operating 128 motor buses and four trailers.[32] Of the some seventy-five thousand square feet of exhibit space provided at the 1924 convention of the American Electric Railway Association, some 80 percent was occupied by bus manufacturers.[33]

Buses enabled traction companies to extend service without constructing new railroad rights-of-way, something absolutely vital as ridership declined after World War I. In 1924, for example, the Indianapolis & Cincinnati Traction Company contracted with two independent bus operators to extend service from Rushville to Brookville and from Greensburg to Versailles, all in southeastern Indiana. The bus companies brought an additional three hundred passengers to the railroad in their first month of operation.[34] Other traction companies operated their own buses, usually through the establishment of a corporate subsidiary. In 1915 the Washington Auto Bus Company, a subsidiary of the Puget Sound Traction, Light & Power Company, inaugurated service from Seattle to nearby Bothell, a community not previously reached by either a railroad or an interurban line. According to Electric Railway Journal, "The cars use a 'street car' body, built by the J. G. Brill Company mounted on an International Motor Company's 2-ton 'Mack' truck chassis. The seating capacity is twenty passengers, with six non-reversing cross-seats, two longitudinal seats of two passengers each, and one four-passenger seat extending the width of the body in the rear."[35]

The use of buses to replace rather than extend operations accelerated after 1925. As journalist Charles Thompson editorialized in the Electric Railway Journal, when an electric line, being burdened by low population densities and thus decreasing traffic, taxes, increasing labor costs, and so on, turned to rebuilding tracks and renewing equipment, abandoning electric lines and replacing them with bus service made sense.[36] The editors of the Electric Railway Journal posed in 1924 a hypothetical but not unrealistic, case: "An interurban electric line some 30 miles long connected two cities, the smaller of which has about 100,000 inhabitants. The railway was equipped with cars suitable for city service, geared for a maximum speed of only about 30 m.p.h. The power supply was poor. The d.c. power plant was nearly 20 miles away from one end of the line with feeders far too small for economic operation. In fact, the voltage drop was so great that on days of heavy traffic, when extra cars were run to handle the peaks, the cars would scarcely crawl up the steeper hills." As the line ran beside a highway, but fully separate from it, there remained an opportunity for relatively fast running, but with stops every few hundred feet, it was not surprising that the cars took nearly three hours to go from terminal to terminal. "Naturally, when the automobile came, traffic fell off until it was deemed wise to abandon through service, and even to tear up portions of the track. . . . A bus line was started."[37] What to do?

But industry trade journals did not need hypothetical cases. News notices frequently appeared announcing line abandonments, with services to be replaced by buses. In 1924 a feature article outlined how and why the Indiana, Columbus & Eastern Traction Company placed in operation twenty-six Studebaker seven-passenger touring cars, thereby offering hourly service between Columbus and Zanesville in Ohio. It was to complement and not replace train service, as the schedules of each operation were staggered. The company added buses, it said, to "control" the highway that paralleled its interurban line, thus to discourage some other motor bus operation from operating there.[38] *Bus Transportation,* the new trade journal for the bus industry, also reported interurban line abandonments and bus service replacements. For example, when in 1932 the East St. Louis & Suburban Railway was abandoned, the journal noted, "Blue Goose Motor Coach Company has taken over the St. Louis–Belleville service entirely, running a bus every ten minutes during rush hours and every twenty minutes at other periods."[39] Between 1917 and 1927, traction company bus routes in the United States totaled some twenty-four hundred miles. During the same decade, passenger service was ended on some twenty-three hundred miles of interurban line. Of that mileage, fifteen hundred miles had been abandoned by 110 companies, which no longer operated at all.[40]

One of the earliest electric railway companies to adopt buses was the Pacific Electric Railroad. The company was formed in 1895 through the merger of the Pasadena & Pacific Railway and the Los Angeles Pacific Railway, becoming a subsidiary of the Southern Pacific Railroad in 1901. The company operated streetcar lines in Los Angeles and other cities, but was also southern California's largest operator of interurban cars with over two thousand trains operating daily over some eleven hundred miles of track.[41] The system was noted for and is today remembered for its distinctive red cars. In 1916 the company began "experimenting" with motor buses, thus to feed passengers to its electric lines. They were assigned to "outlying districts which it is desired to connect with existing traction lines or where it is not practicable to build extension under present conditions," it was reported. The "buses" were actually trailers hooked to modified automobiles.[42] In 1927, 114 buses were being operated in intracity service, making for some 4.2 million car miles annually and carrying nearly 1 million passengers a month.[43] In 1931, the company was operating over four hundred buses.[44] Pacific Electric, was one of the nation's more viable interurban operators. Nationwide between 1917 and 1930, vehicle miles for interurban cars held constant, but for buses rapid growth occurred, especially after 1924. Pacific Electric was one of the companies leading the way.

MOTOR TRUCKS

With passenger revenues either stagnant or declining, interurban managers turned increasingly to freight traffic for improved profitability. The Detroit United Railway was organized in 1900 through the consolidation of several traction lines serving the city. From the very beginning, it used combination passenger and baggage cars to deliver packages along its lines. Package freight was delivered to wagons and then trucks in small towns and villages, usually in its main street. If no pickup had been arranged, then it was merely stacked at curbside for consignee retrieval. Loading and unloading cars at mid-street not only delayed trains but also created traffic congestion. In 1912 the company introduced box motors, most of them converted from discarded stripped-down passenger equipment, that were capable of running on their own schedules, usually at night after passenger service ceased. Then came freight trailers to create true multicar trains. In Detroit a large freight house was constructed near the Detroit River, suitable not only for convenient truck pickup and delivery in the city's downtown, but close to the docks, with freight being exchanged in the summer months with Great Lakes passenger steamers. The railroad developed an extensive traffic in milk, with special trains operating on all of its divisions. In 1914, some two thousand cans of milk were brought into Detroit each day, shipped from distances as far as thirty miles away.[45]

Many traction companies initially embraced trucking by contracting with independent truckers to provide what was called "door-to-door service." Interurban lines carried mainly package freight (or "less-than-carload" freight). Truckers at one end picked up customer shipments and deposited them with the railroad. Truckers at the other end picked them up and delivered them to consignees. Thus *Electric Railway News* reported in 1916: "Articles of incorporation have been filed by the Mount Eden Motor Truck Company, which will operate the truck end of the joint freight service with the Louisville & Interurban Company." The line extended east from Louisville to Shelbyville, with the trucking company operating at both locations.[46] As highways improved, however, truckers began establishing their own long-distance routes to compete directly with rather than complement the interurbans. The next step for the traction companies was to develop their own trucking services through the establishment of truck subsidiaries.

It was not surprising, journalist Charles J. Laney argued in 1918, that motor trucks had entered the transportation field, and not just for local deliveries but for long-distance carrying also. The interurbans needed to respond. Most had been slow to do so, much to their own detriment. "Motor-truck transportation had its start in territory served by interurban lines," Laney wrote. "These lines were built at a great cost. Rates were made very low and frequent service was established. Real estate developments in the suburbs were encouraged. Excursions were run at frequent intervals, and inter-community intercourse was encouraged in every possible way." But there had been relatively little thought of freight. Now truckers seemed to have everything their way. Rural roads were being improved at public expense. Truckers had their rights-of-way provided and maintained through public tax monies.[47] Why wouldn't business interests want to establish trucking companies? But couldn't the interurban lines still compete? There were, in fact, many factors why many traction companies did not. Some companies could not carry freight, given the franchise restrictions under which they were licensed. Many traction rights-of-way, with their sharp curves, steep grades and light rail, could not handle trains with multiple cars. Nearly every interurban line suffered from an insufficient number of long, double-ended sidings whereby passenger and freight trains could readily pass. They lacked adequate freight houses and freight yards, and, above all, they needed more and better rolling stock. Truck technology was rapidly advancing. Competition from truckers would only be getting worse.

Trucks, journalist A. B. Cole pointed out in 1918, were proving themselves on the front lines of France: "There, where the railroad facilities are generally inadequate and where results must be obtained, regardless of cost, the motor truck has been wonderfully effective in affording a quick, though costly, means of transportation over rough roads." What worried Cole was the legislation then pending in Congress for "the maintenance, repair and reconstruction of public roads." Eventually the federal government would be wading in waist deep to build and maintain a system of interstate highways, with national defense being one of the reasons for doing so. It would be a waste of money, he argued. "If the regular avenues of transportation, viz: the railroads both steam and electric, were used, this unnecessary expense account of repairs to roads due to movement of . . . heavy trucks over the same could be saved."[48] But, of course, the railroads were failing to keep up with war-imposed transport demands. Only a few interurban lines were positioned to help out.

Interurban rights-of-way that were well separated from rural highways and built more to steam railroad standards—for example, well provided with passing tracks and even bypass lines around congested business districts—were better suited for freight haulage as they were better suited to heavy traffic generally. Beginning in 1926, the Chicago, Milwaukee & North Shore offered "Ferry Truck Service" over its eighty-five miles between the Chicago Loop and downtown Milwaukee. It was nothing less than a pioneering "piggyback" service, with small trucks initially and later large truck trailers being carried on flat cars. Customers delivered freight to off-line freight stations located throughout Chicago and Milwaukee. It was then transferred to the railroad in traction company trucks. The piggyback formula was suggested when customers were capable of filling entire trucks or entire truck trailers.[49] Then company vehicles were loaded at customer loading docks and then sent on flat cars to be delivered directly to consignee locations. Indeed, it was the traction industry that pioneered piggyback railroad service.

In the 1920s, several interurban companies pioneered container shipping, including the Detroit United Lines in Michigan and the Cincinnati, Aurora & Lawrenceburg Railway in Ohio and Indiana. Offered was "forwarding under seal direct from shipper to consignee, thereby economizing in station platform labor and reducing claims for pilferage and damage; lessening chance for shipments to go astray, and relieving congestions at freight stations."[50] Only a relatively few interurban lines moved commodities in bulk, operations that required powerful electric engines, heavy cars, and thus heavy track. Pictured is a load of scrap iron on the Youngstown & Ohio River Railroad (fig. 6.8). Transport of heavy bulk commodities by railroad, even traction railroads, was superior to trucking cost-wise and perhaps even time-wise. The Terre Haute, Indianapolis & Eastern introduced a fleet of stock cars in 1921. In 1923 it and other traction lines transported some 11,500 loads of cattle, sheep, and swine into the Indianapolis Stock Yards.[51]

Pacific Electric maintained a large freight service. The railroad had struggled for many years to meet truck competition, lowering freight rates to do so, only to be stymied by lower rates from truckers. Thus was a trucking subsidiary, the

FIGURE 6.8. Youngstown & Ohio River freight train, 1924. T. H. Stoffel, "Freight Service," *Electric Railway Journal* 64 (Sept. 27, 1924): 506.

Pacific Motor Transport Company, organized in 1929. The *Electric Railway Journal* reported that its charter gave it the right to "engage in business as an express company, as a common carrier by motor truck on the highways, as a local city drayman and as a warehouse company." It could, in other words, do everything a trucker might do. Three years later the subsidiary was operating over some six hundred miles of route, serving "almost every important municipality in Southern California." Initially freight was funneled to some twenty truck-convenient freight stations located on various branches of the Pacific Electric system. In 1931 the number had been enlarged to over two hundred, but located on both the Pacific Electric and its parent, the Southern Pacific Railroad.[52]

Having reached a peak of some 15,500 miles in 1918, interurban mileage declined to some 3,100 miles by 1940.[53] Today, less than two hundred miles of former interurban miles remain, most of that trackage operated by a single railroad, the Chicago, South Shore & South Bend, itself long a subsidiary of the Chesapeake & Ohio Railroad. It is clear what accounted for the industry's decline. First, most interurban railroads were quickly and thus lightly (if not shoddily) built, and for local passenger service only. Second, anticipated patronage on most lines did not materialize. In consequence, interurban companies could not finance needed upgrading as tracks and rolling stock deteriorated. But clearly more than anything it was the coming of the motorcar, motor bus, and motor truck that sank the industry. Replacing passenger service with buses and adopting trucks to improve the pickup and delivery of freight offered but temporary relief. What started out in part as an expression of the nation's Good Roads Movement— improving rural areas with rails laid along highways—ended when rural highways instead were fully hard surfaced. The handwriting was clearly on the wall as early as 1916. As the general manager of one Upstate New York interurban line wrote, "The real competition is the privately owned automobile, and we must assume that the privately owned automobile will continue to increase in numbers. More and better brick, concrete and asphalt roads are being constructed, and the price of automobiles continues to decrease—factors which inevitably spell 'more automobiles.'"[54] We end this chapter with a photo taken in the small Illinois town of Ridge Farm in about 1916 (fig. 6.9). An interurban car of William B. McKinley's Illinois Traction Company scoots past a residence. An automobile sits at the curb. The owner of the house sits at the wheel. With interurban cars passing by his front door, he prefers to drive.

FIGURE 6.9. Automobile meets interurban, Ridge Farm, Illinois, circa 1910. Photograph, authors' collection.

Chapter 7

RAIL MOTOR CARS

As new interurban lines began to syphon off passengers, the steam railroads began to experiment with self-propelled passenger coaches similar to those of the traction companies except that power was not drawn off overhead trolley wires. Little effort was made to apply steam power to such vehicles even though some automakers were themselves producing steam cars. A penchant for steam undoubtedly lay with the traditional railroader's fondness for steam locomotives, but it was the internal combustion engine that won out. At first, power was transmitted mechanically to car trucks (axel and wheel sets) through transmissions which, although like those of motor trucks, were much heavier, given the added weight of the vehicles. Engines, on the other hand, were sized very much like those of motor trucks. However, internal combustion engines were quickly combined with electrical motors and the motors then used to actually turn car wheels. Initially, small bus-like vehicles had been adopted to ride the rails. But full-sized self-propelled coaches soon emerged, vehicles that looked more like interurban cars if not standard railroad coaches.

REPLACING STEAM LOCOMOTIVES

With traction lines syphoning off short-distance passengers, railroad officials debated how they might retain such patronage, if not tap into additional ridership by emulating interurban service. Primarily they were concerned with reducing the costs of operating short-distance passenger trains equipped with standard steam locomotives and rolling stock. As we elaborate below, the cost of running a train pulled by a steam locomotive, even with only a single passenger car, was substantially higher than that of running a single self-propelled car as the traction companies did. Additionally, with costs cut, self-propelled cars could run more frequently. The *Railroad Gazette* editorialized in 1906: "In the early days of interurban electric competition, the steam roads sought to win back local traffic by cutting rates, but this procedure was soon found to be ineffective. Most steam road managers now realize, or should realize, that the most attractive features of the electric road are frequent service, high speed, and accessibility to the passenger, and unless the steam road can offer these same advantages, the bulk of the car traffic will continue to go to the interurban roads which use electric motor cars." Experiments were made to validate whether railroads could not duplicate what the interurbans offered, but, of course, "without the necessity of a large investment in electrical apparatus, overhead or third-rail construction, motor cars, etc."[1] At the 1903 Railroad Congress held at Paris, one working session dealt with that idea. The session was titled, "Adoption of Automotor Cars (Steam, Petroleum, Electricity) for Working Normal Gauge Lines With Light Traffic."

Already cars were being readied for trials both on the Baltimore & Ohio and the Great Northern.[2] But most of the important experimenting had already been done in Europe, where rail motor cars were already in service, especially in England, Germany, Austria, and Hungary. One of the first notices that actual rail motor car service had begun in the United States appeared in 1903 in the *Street Railway Journal*. An Iowa short-line railroad, the Tabor & Northern Railway, which connected Tabor, a small college town of some one thousand people, with the main line of the Chicago, Burlington & Quincy about nine miles distant, had put into service "a gasoline railway car" similar to the "auto coaches which are such familiar sights in Chicago streets." The report continued, "The entrance is at the front, so that all passengers in and out of the car pass the motorman, who collects the fares as well as operates the car, thus dispensing with the services of a conductor. As this car is considerably lighter than a light interurban car it can be operated over a much lighter track without excessive track deterioration." Its two-cylinder engine produced 25 horsepower and had a top speed of twenty miles per hour. The cost of gasoline was calculated to be one cent to the mile.[3]

The *Railroad Gazette* reported in 1906 that "gasoline railroad cars" had been put in service by several "prominent railroad companies." It elaborated, "On branch lines in isolated districts, where the service is intermittent, the cost of maintenance and operation of a locomotive and one or two cars is far in excess of the revenue, and the gasoline motor car is certain to make such lines remunerative, the labor being materially reduced and the fuel consumption being proportionate to the work performed." But the "motor car proposition" was meeting "its share of antagonism," the report went on. "In its mildest form this appears in the skepticism expressed by the older type of railroad man, who has 'run the gauntlet' in the development of the present steam locomotive, and who does not believe that any vest-pocket motor can share in the same class of work. In his opinion, the gasoline motor may be all right as a necessary adjunct to the automobile, which is a toy and a luxury at best, but that it can ever legitimately undertake the performance of real hard railroad work at a profit is dubious in the extreme."[4]

By 1907, however, opinions were quite different. William R. McKeen Jr., superintendent of power for the Union Pacific Railroad argued in a speech delivered to the New York Railroad Club that steam locomotives were not only inappropriate for branch-line operation, but steam-powered rail motor cars were also: "The modern locomotive and steam motor car, with high steam pressure and the attendant flue and firebox troubles—the troubles due to formation of scale, broken staybolts, leaky front-ends, defective draft, poor coal and kindred necessary evils incident upon the use of a separate power-generating unit, such as a boiler entails—are much more complicated and vulnerable pieces of machinery than the gasoline motor car in which, technically speaking, there is nothing present but 1) vehicle, 2) 'prime mover,' and 3) transmission; the complicated generator with its attendant parts likely to give trouble being absent here." He concluded, "Branch lines collect freight traffic and feed the main line, and the limited passenger business, of course, can be handled economically when turned over to the main line. Thus it is, if the steam train could be replaced by a combination motor car, a great saving could be made in operating expenses. Passenger traffic which would be insufficient to fill a steam train in most cases would justify the operation of a gasoline motor car."[5]

MCKEEN MOTOR CARS

In 1905 the Union Pacific Railroad introduced its first gasoline motor car under McKeen's direction. His father was a banker and president of the Terre Haute & Indianapolis Railroad (dubbed the "Vandalia" after it was extended westward by way of Vandalia, Illinois). It was the line that the Pennsylvania Railroad used to complete its Panhandle Division (which ran west from Pittsburgh to Columbus and Indianapolis) on to St. Louis. The Vandalia's large shop facility at Terre Haute was for many years that city's largest employer. Educated at Terre Haute's Rose Polytechnic Institute, Johns Hopkins University, and the University of Berlin, the younger McKeen took over supervision of those shops, and then moved on to the Union Pacific Railroad, first at North Platte, Nebraska, and finally at Omaha. Edward H. Harriman, owner of the Southern Pacific and the Illinois Central among other railroads, had bought a very much dilapidated Union Pacific only to totally revitalize the company through track replacement and new motive power and new rolling stock, making it into one of the nation's most modern railroad operations. He encouraged McKeen to experiment with new kinds of railroad equipment, which produced the first truly viable rail motor car.

The 1905 model was fully experimental (fig. 7.1). Its design mimicked that of a racing yacht, the *Railway Gazette* reported. "The rear of the car is rounded off to avoid the formation of the vacuum present with square-ended cars. The front tapers to a sharp point. The roof is given a taper from the middle to produce a splitting effect in the atmosphere and minimize resistance."[6] It would eventually be recognized as one of the earliest examples of "streamlining." The car seated twenty-five passengers. The internal combustion engine was built by the Standard Motor Works of Jersey City, New Jersey, a manufacturer of engines for boats. When completed, the car was run from Omaha to Portland, Oregon, where it went into regular service. Gasoline consumption on the trip averaged three quarts a mile.[7] Within the year the Union Pacific had built six more cars, each with larger six-cylinder, 200-horsepower engines. Then, ten even larger seventy-five-passenger motor cars went into production. In 1907 the *World's Work* took note: "Day in and day out on branch lines of the Union Pacific Railroad near Omaha, and at other places, sleek steel cars, like sharp-nosed torpedo boats on wheels, slip along at from forty to sixty miles an hour with passengers gazing out through air-tight windows into a smokeless atmosphere."[8]

Gasoline engines powered electric motors by which the cars were actually moved. Some of the later cars transmitted power mechanically to the car trucks

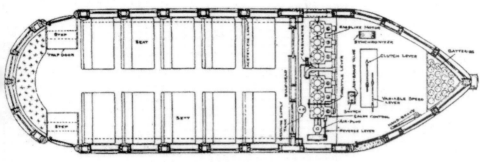

FIGURE 7.1. Union Pacific Railroad Experimental Motor Car Number One, 1905. "The Union Pacific Gasolene Motor Car," *Railroad Gazette*, n.s., 39 (July 7, 1905): 14.

through a rugged transmission similar to those used on motor trucks. "It is ready to move at once without any waiting to get up fires, is indeed practically an automobile on rails." The cars could be run by two employees, a motorman and an attendant. Although they used gasoline (at nine cents a gallon), cost of running the cars compared very favorably with that of steam trains. "A steam train with locomotive and two cars (with such expenses as labor, repairs, and cleaning included) cost twenty-four cents a mile to operate," the article continued. Electric service with one motor-car and a trailer cost eighteen cents, but also necessitated expensive trolley-wires or third-rails. On the other hand, a gasoline motor-car with a trailer, offering baggage, mail, and express as well as passenger service, costs 15 cents a mile.[9]

The McKeen cars were built in the Union Pacific's shops at Omaha. In 1908, however, an independent McKeen Motor Car Company was organized, with the railroad and McKeen each owning half. The railroad, having just completed new shop facilities, turned the old buildings over to the new company. Pictured is the company's standard car produced over the next several years (fig. 7.2). The *Railway Age Gazette* reported:

> The circular or port-hole windows, which are a peculiar feature, and which give an almost unobstructed panoramic view from the inside, enable the sides of the car to be used as a deep girder in the design of the framing, with over 8 feet between top and bottom chord. In addition to permitting this great increase in the strength of the car side, these windows are said to be greatly appreciated by passengers on account of their tightness when closed. Even the fine dust of

the western prairies is excluded, as well as winds and rain, by the rubber sealing gaskets. The tapered front end, by decreasing wind resistance and facilitating acceleration, becomes an economical feature, lessening the consumption of gasoline per mile run. The side door entrance is another peculiarity. The roof is lower than that of the ordinary passenger car and the width is greater. The special ventilating system affords a change of air every four minutes, if desired. The cars are built of steel.[10]

The company's cars were sold to other railroads, as well as employed by the Union Pacific. For example, in 1909 a custom built car was sold to the Norfolk & Southern Railroad, which connected Norfolk in Virginia with Charlotte in North Carolina. "It contains a 7-ft. railway post office, a baggage and express room, a compartment for colored passengers seating 10 persons, and a compartment for white passengers seating 50 persons," noted the *Railroad Gazette*. "The compartment for white persons will be divided into a smoking compartment, seating eight, and a regular compartment seating forty-two. The seats will be rattan."[11]

John H. White, writing in his classic two-volume work, *The American Passenger Car*, noted that, despite the McKeen company's relatively short life, its products left "an imposing legacy." Of the 150 cars built, many were still running in the 1930s. More important, he said, they ushered in "a new era of rail technology" that influenced many of those who in the 1930s would bring to the fore modern streamlined passenger trains. Edward G. Budd, one of the foremost passenger-car builders before and after World War II, "cut his teeth," White em-

FIGURE 7.2. Advertisement for the McKeen Motor Car Company. *Railway Age Gazette*, n.s., 57 (Dec. 25, 1914): 18.

phasized, "working on McKeens." Budd freely conceded that early in the twentieth century Harriman and McKeen had "foreseen the engineering principles"
embodied in "Zephyr type" passenger trains.[12]

OTHER MANUFACTURERS

The McKeen cars were something of an anomaly given their size. A self-
propelled motor car with baggage and passenger compartments, and with a trailer
in addition, was very much train-like. Most other motor cars looked and were
more like interurban cars or even buses. Many companies were involved in the
production of rail motor cars, a few assembling the cars and many others supplying components, especially the gasoline engines and electric motors. Several
railroads—for example, the Union Pacific—engaged in rail car construction at
their shops. But most cars were purchased. Technology was far from uniform
and, early on, it was indeed quite experimental. The *Railroad Gazette* reported
in 1906, "Apparently the greatest difficulty encountered by the designers of the
gasoline cars is the transmission of power from the engine to the driving wheels.
This is not surprising in view of the fact that the internal combustion engine is
essentially a constant-speed motor and that railway work demands wide ranges
of variable speeds. At the present time the favorite means appears to be the use
of electricity, indicating that the difficulties of direct mechanical transmission
and variable speed operation are so great as to warrant the rather roundabout
transmission involved by addition of a generator, with or without batteries, and
the standard railway type motors and control."[13]

The Stover Motor Car Company of Freeport, Illinois, introduced a thirty-
five-passenger rail motor cars intended for "use on smaller branch lines, or on
logging roads, where it is desired to handle a passenger business." It looked very
much like a motor bus, its four-cylinder, 30-horsepower gasoline engine being
located in the front under a truck- or bus-like hood for ease of accessibility (fig.
7.3). Being lightweight, the vehicle had a standard automobile transmission to
convey power to the rear wheels. It followed from an earlier vehicle: a twelve-
passenger, open-sided automobile with flanged metal wheels.[14] In 1909 the Stover
company was sold to the Buda Foundry & Manufacturing Company of Chicago,
makers of railroad inspection and section cars as well as "velocipedes" (or self-
propelled handcars). In 1906 the J. G. Brill Company of Philadelphia developed
a gas-electric motor car for the Missouri & Kansas Interurban Railway, a line

owned by William B. Strang that connected Kansas City with Olathe, Kansas. Over ensuing years, the "Strang Car" was subsequently marketed by the Strang Electric Railway Company to a number of steam railroads, starting with the Chicago & Alton and the Toledo, St. Louis & Western (the Cloverleaf Line). According to the *Railroad Gazette*, "The propelling apparatus consists of a six-cylinder gasoline engine direct[ly] connected to a direct-current dynamo in the engine compartment, a starting rheostat, storage battery mounted under the car, a series-parallel controller and two motors on the axles."[15] It looked very much like a streetcar (fig. 7.4).

The J. G. Brill Company, for its part, actually went on to become one of the nation's largest manufacturers of streetcars and, between 1920 and 1930, its largest maker of heavier electric interurban cars. But the company also made car bodies for other manufacturers—for example, the International Motor Company's "one-man" rail motor car introduced in 1921 and initially put in service on the Chesapeake-Western Railway in Virginia. "The car resembles the large type of single-deck motor bus extensively operated on the streets of cities," noted *Electric Railway Journal*, "with the difference that under the front of the vehicle is a small four-wheel pivotal truck which enables the vehicle to take curves more easily and at higher rates of speed. . . . The car consists of a 2-ton Mack dual-reduction motor truck, a Mack standard engine, and transmission."[16] Other firms that assembled rail motor cars included the Bowen Motor Railways Corporation of St. Louis, the Smalley Rail Car Company of Davenport, Iowa, the Russell Car

FIGURE 7.3. Stover rail motor car. "The Stover Motor Car," *Railway Age*, 44 (Nov. 22, 1907): 750.

FIGURE 7.4. Strang gasoline-electric rail motor car, *Railroad Gazette*, n.s., 40 (Feb. 23, 1906): 189.

Company of Ridgeway, Pennsylvania, and (with the most unusual name) the Four Wheel Drive Auto Company of Clintonville, Wisconsin.

Auto and truck manufacturers, such as the Stover Motor Company, had helped prompt rail car production. The gasoline-propelled rail passenger car, claimed *Railway Age,* had had its commercial inception through the desire of truck manufacturers to increase their production, as sales of motor trucks, for example, had fallen off following the close of World War I. "The adaption of the standard motor truck chassis to rail operation immediately directed the attention of the gasoline motor to railway cars," noted *Electric Railway Journal,* "but the demand was almost immediate for units of larger capacity than the truck manufacturers cared to furnish. The 'child' quickly outgrew its parental home."[17] After 1920, railroad car builders quickly came to the fore, although auto and truck makers still supplied engines among other equipment.

RAIL MOTOR CARS IN OPERATION

By 1920 the rail motor car had become as common a sight as electric interurban cars and were probably more widespread, given that they were employed throughout the United States and not concentrated regionally. They replaced steam trains on unprofitable branch lines and were placed in seasonal service in resort areas. Pictured is a rail car and trailer standing at the ready, interchanging passengers with a steamer at Coeur d'Alene, Idaho (fig. 7.5). Almost everywhere motor cars proved to be cost savers, but not always. In 1909, a motor car was introduced between Salinas and Herrington in Kansas, a distance of some fifty miles, on the Chicago, Rock Island & Pacific. The average operating cost in November was 15 cents per mile, 18.5 cents in December, and 20.4 cents early in the following January. Then it was replaced late in January by a steam locomotive and one passenger car at a per-mile cost of 17.3 cents. Clearly that rail motor car was, at best, a warm-season alternative.[18]

One problem was the steady increase in the cost of gasoline as motorist demand for the fuel pushed prices up. There were, however, alternative fuels on the horizon. Should the cost of gasoline exceed twenty cents a gallon, another motor fuel just as good could be more cheaply produced from vegetable material. But then there was also the new diesel technology being developed in Germany. Perhaps it would be possible, speculated one journalist, to design a railcar or even a locomotive high in efficiency, light in weight, and fully rugged to provide reliable service with ease of operation and ease of maintenance.[19] But instead the price of gasoline plummeted in the 1920s with the opening of new oil fields and larger and more advanced oil refineries across the country. And it was not until

FIGURE 7.5. Rail motor car at Coeur d'Alene, Idaho, circa 1920. Postcard, authors' collection.

1925 that the first diesel locomotive was placed in regular service in the United States on a passenger train operated by the Central Railroad Company of New Jersey.[20] Through the remainder of the 1920s and into the 1930s, diesel locomotives were used mainly as switch engines in freight yards.

The New York Public Service Commission issued a report in 1915 sounding a warning to the steam railroads of the state. As the *Electric Railroad Journal* editorialized, "They must find some more convenient and economical means for handling local passenger business to meet the competition of the electric railway and the automobile unless they want to see a continuance of the falling off in local passenger revenue." Thus the commission urged the railroads to further "study the possibilities of the gasoline-driven railroad car." The journal continued, "The familiar local train, composed of locomotive, baggage and express car, and two or three coaches, is in a state of obsolescence. Such trains carry only those who have no other available means of transportation. Some cheaper, faster and more convenient method of transporting local passengers must be adopted or else the railroads must continue to transport them at a loss in spite of reasonable curtailments of service." The automobile was the culprit: "A few years ago the summer toy of the rich man, it has now become the convenient passenger and freight vehicle of almost all classes. It is safe to assume that its use will not diminish."[21]

In 1922 the New York, New Haven & Hartford Railroad compared the performance of a gasoline rail car and a motor bus between New York City and New Haven. Rail cars made the sixty-seven-mile trip in an average of two hours and forty minutes at a speed averaging twenty-five miles per hour. Buses took forty-two minutes longer over seventy miles of city street and rural highway at an average speed of twenty-one miles per hour. The rail cars carried thirty-five passengers. The buses seated twenty-five. Costs of operation were not calculated.[22] Time-wise, at least, the rail motor cars had proven advantageous. Such advantage, however, would substantially vary from locale to locale, a function of street and highway traffic volumes, but also the nature of a railroad's right-of-way and equipment.

There were many variables. "When an engineering officer of a railroad undertakes the selection or design of a unit rail-car," wrote journalist W. L. Bean, "he must consider carefully the factors of size of car, speed, grades, curvature, power of engine, transmission and control system, gear-ratio, heating, lighting and seating arrangement, if he is to secure a satisfactory vehicle." The engine, he presumed, would use gasoline "since Diesel-type engines" he said, "have not been produced and time-tried in light weights at horsepower ratings desired in passenger rail cars."[23] The Philadelphia & Reading Railroad did calculate costs, running both a gasoline-electric railcar with trailer and a steam locomotive with one passenger coach over the same forty-mile route in 1928. The railcar operation showed a net savings of some $1,500 per month or some $18,800 per year. Considered in the equation were not only fuel costs but also crew wages and repairs. Railcar advantage meant that 30 percent of its cost of purchase might be returned in its first year of operation.[24]

Fifty-one motor cars were ordered by steam railroads in 1922, 76 in 1923, 120 in 1924, and 149 in 1925.[25] Among the largest purchases were those made by New England railroads, not only the New Haven but also the Boston & Maine and its subsidiary, the Maine Central. Besides increasing in numbers, railcars were becoming heavier and somewhat more powerful, although engines seldom exceeded six cylinders with 100 horsepower. In 1930 the New Haven, with an unusually large number of short branch lines, had substituted rail motor cars for thirty-six steam locomotives and seventy-two antiquated wooden coaches, saving in the process operating charges of approximately 50 percent.[26] But the railroad had also established a bus subsidiary, the New England Transportation Company, to provide service along highways that closely paralleled its lines.

Important was the fleet of sixty rail motor cars operated in 1931 by Chicago, Burlington & Quincy Railroad over some thirty-three hundred miles of track largely in the Midwest. Almost 60 percent of the mileage involved one or two trailers. Cost savings were quite clear. According to *Railway Age,* "The cost of gasoline and lubricating oils per mile was 6.11 cents; repairs, labor, and material, 5.03 cents; and total operating expense, 27.04 cents. Allowing 6 per cent interest on investment and a generous depreciation rate of 8 per cent, these gas-electrics . . . earned 28.5 per cent on the investment in 1930, and saved $699,290, or 36.3 per cent of the cost of equivalent steam train service."[27] Not only were steam locomotives retired, but their retirement also enabled the closure of coal, water, and sand-and-cinder-handling facilities. Retired locomotives were on average forty years old, reflecting the custom of the railroads to use their oldest equipment on short branch-line runs. The railroad's first gas-electric cars had gone into service in 1927, but they were so reliable and inexpensive to operate that the railroad quickly introduced them on its main lines as well. The Atchison, Topeka & Santa Fe placed six heavy-duty rail motor cars in service in 1931 with

500-horsepower engines built by General Electric. The year 1928 had seen a big step toward higher-powered units nationwide, with over 70 percent of new rail-cars thereafter having power plants exceeding 350 horsepower. In 1931, however, only thirty rail motor cars were ordered nationwide. For the first time in a decade no trailer cars were ordered.[28] The Great Depression had set in.

The day of the self-propelled railroad motor car was coming to an end. Fewer and fewer new vehicles were purchased each year. That is not to say, however, that some of those in service did not continue through to the 1940s and even the 1950s. A kind of nostalgia for surviving rail motor cars evolved among railroad enthusiasts after World War II, the term "doodlebug" being coined for the purpose.[29] However, substantial branch-line motor-car service had already been abandoned by the start of World War II. Falling passenger traffic simply made continued service impossible on most branch lines. And, as we will see in chapter 8, railroad bus subsidiaries were also trimming rail service. Americans preferred traveling by car, certainly over short distances. They clearly did not prefer dated rail motor cars any more than dated interurban cars—nor, for that matter, the new buses. Transit simply lacked the convenience and speed of motoring by automobile. Buses would remain important for travelers between the nation's largest cities, but mainly for those who did not own cars. Yet the rail motor car did not entirely go away. It was rejuvenated for long-distance travel in a new form. The self-propelled railroad motor car became the prototype for the modern streamlined passenger train.

STREAMLINING

Today it is remembered that the streamlining of railroad passenger trains derived from aircraft designs intended to diminish air drag and increase flight efficiency. Largely forgotten is that the new streamlining was not applied initially to steam locomotives and passenger cars, but first to self-propelled rail motor cars and their trailers. Indeed, streamlining at first involved only single self-propelled rail cars. But it was quickly applied to integrated or "articulated" motor cars and trailers, thus to create what were then called "train sets." It was the established railroad motor-car manufacturers who began the trend. But the nation's largest railroad car builders quickly joined in, among them firms like the Pullman Car & Manufacturing Company, which in 1934 merged with the Standard Steel Car Company to form Pullman-Standard.

In the early 1930s, the Four Wheel Drive Auto Company introduced a bus-like vehicle with rounded corners and flowing horizontal lines fully suggestive of design modernism. It was very much a modern bus on rails. It featured a three-speed transmission, four-wheel drive, and rubber wheels, which were held on the rails by metal flanges. It seated thirty-five passengers in the day-coach version while a sleeper model had sixteen berths. The company called it their "rail bus."[30] Philadelphia's Edward G. Budd Manufacturing Company became the leading advocate of streamline-modern railroad equipment, utilizing stainless steel, aluminum, and other materials seldom before seen on railroad rolling stock. A self-propelled rail car with "a body of extremely light weight for use in light branch-line service" was delivered to the Reading Railroad, *Railway Age* reported. "In this design the car-body load was carried on a trussed underframe structure with sections built-up of thin high-chromium-steel sheets of stainless quality, using . . . spot welding" (fig. 7.6). Its exterior was paneled in stainless steel and thus it did not require a protective coating against corrosion. It rode on rubber tires fitted with metal flanges, which the Michelin Company had innovated in France for railroad service there. It was some fifty feet long and could carry forty-seven passengers. The car weighed a mere 468 pounds per seated passenger. Importantly, it was propelled by a single 125-horsepower diesel engine. The Budd Company estimated its operating cost at some 12 cents per mile.[31]

At the same time the Pullman Car & Manufacturing Company developed its own lightweight, high-speed rail car called the "Railplane." It was sixty feet long and could carry fifty passengers at a top operating speed estimated at ninety miles per hour. "It is featured by welded tubular steel construction for both the body and the truck frames; an outside covering of aluminum alloy sheets, conforming to the streamlined contour, and two driving motors mounted directly on the front trucks," noted *Railway Age*. The car's shape and styling were based on automotive and aeronautic design precedents created by William Bushnell Stout of the Stout Metal Airplane Division of the Ford Motor Company in Dearborn, Michigan. "The fundamental objectives sought in the design were adequate strength with the least possible weight to decrease rolling resistance, full streamlining to reduce air resistance at high speeds and, consequently, low power requirements and economy of operation." The framework of the car was of chrome-molybdenum steel tubing, noted for its high tensile strength and resistance to corrosion. "The tubular material was readily welded, thereby forming an integral structure which has proved exceptionally satisfactory in air-craft work." Thus the Ford Motor

FIGURE 7.6. Budd Company rail motor car for the Reading Railroad. "Budd-Michelin Car Delivered to the Reading," *Railway Age*, n.s., 93 (Nov. 12, 1932): 669.

Company was said to be "dabbling in air-minded railroading," when the first car was placed in service on the Gulf, Mobile & Northern Railroad in Mississippi.[32]

Early on, the Seaboard Airline Railroad was among several steam railroads to adopt streamlined rail motor cars. Beginning in 1935, a "rail-bus motor coach train" ran between Richmond and Raleigh on the railroad's main line, thus "to relieve the through steam trains of local stops." Service involved eight regular and twenty-two flag stops during some four hours of transit along a 314-mile route. It was built by the American Car & Foundry Company at their St. Charles, Missouri, works and was one of first railroad passenger cars to be air conditioned.[33]

EARLY STREAMLINERS

The Union Pacific continued use of its McKeen motor cars but had also bought rail cars from other manufacturers. Then it commissioned the Pullman Car & Manufacturing Company to design and build a streamlined, diesel-powered train set based on, but substantially different from, William B. Stout's Railplane; it was the Union Pacific M-10000. It was a three-car articulated train, which, once tested, became the City of Salina operating between Kansas City and Salina, Kansas. Then came the six-car M-10001 intended for even longer-distance travel. It was the actual prototype for the many other train sets that followed. It included a power car, a mail-baggage car, three sleepers, and a coach with buffet service. It is pictured here from the cover of a 1935 Union Pacific advertising brochure (fig. 7.7). Aluminum-alloy construction was used throughout the train with the exception of its power plant, bolster frames and sills, and, of course, the trucks (axle/wheel assemblies) all made of steel. The train was 376 feet long, with power being supplied by a 600-horsepower Winton diesel engine (with later trains using 900-horsepower engines). The engine powered four electric traction motors, which then conveyed power to one truck assembly. The Winton Engine Company, recently purchased by General Motors, was the descendant of the Winton Motor Carriage Company, originally a manufacturer of bicycles, and then the maker of one of America's pioneering automobiles.

Streamlined train sets featured articulated cars. That is, they shared their trucks with adjacent cars. Each car, in other words, swiveled on shared trucks placed at each end. With conventional passenger coaches, each car had its own separate trucks, and the cars were being coupled together. Articulation not only reduced the number of trucks needed but also eliminated the need for couplers. And It greatly facilitated passengers moving from car to car since no doorways obstructed passage, cars being open one to another. (There was, however, a serious downside: with articulated cars, trains could not be easily added to or subtracted from in order to fit changing traffic demands.) Sleeping cars included many innovations. For example, a sliding door in an aisle partition afforded lower-berth privacy, making the berth into a kind of "mini-roomette." Upper berths were reached by a foldout ladder and were partitioned from the aisle by a "rolling shutter" (fig. 7.8). Trains were air-conditioned with "Frigidaire cooling" supplied by General Motors' Frigidaire Division.[34] The long-distance trains that followed the City of Salina variously connected Chicago and St. Louis with cities to the west, and were named accordingly: for example, the City of Denver, the City of Los Angeles, the City of Portland, and the City of San Francisco.

The M-10000 had been designed more as a "laboratory on wheels." When it was completed, the train was taken on tour across the United States, traveling over fourteen different railroads in twenty-two different states. According to Railway Age, "It crossed all the important mountain ranges of the country; it was subjected to every sort of weather condition in all altitudes up to 8,000 ft.; it was in blizzards and temperatures as low as 10 deg. Below zero in the east and nearly 100 deg. Above zero on the Pacific coast; it experienced snow, wind, rain and dust storms; it operated continuously on the schedules of the fastest trains on the several railroads over which it passed; it successfully negotiated grades

FIGURE 7.7. Union Pacific Railroad's articulated streamliner.
Advertising brochure for Union Pacific Railroad, circa 1934.

and curves of all railroads." Specialists aboard analyzed the performance of its
gasoline-electric power, the articulation of the cars, air flow over the streamlined
outer shell, the air-conditioning, the braking, and the riding qualities.[35] After
it passed its tests the train was proudly displayed at Chicago's 1933 Century of
Progress Exposition.

The New York, New Haven & Harford's "Comet," built by Goodyear Zeppe-
lin, was the next streamliner to attract attention (fig. 7.9). The New Haven had
been an important early adopter of rail motor cars, given its numerous short-
branch lines variously connecting cities large and small in Connecticut, Rhode
Island, and eastern Massachusetts. The first diesel-powered Comet connected
Boston with Providence forty-four miles away, a run that took only forty-four
minutes. *Railway Age* reported, "Traffic and publicity officers have labored aggressively to bring these trains to the attention of the public; operating officers
have established splendid records of on-time performance with absolute safety.

FIGURE 7.8. Upper and lower berths in Union Pacific's streamliner sleeping cars. "U.P. Gets Second High-Speed Train, *Railway Age*, n.s., 97 (Oct. 13, 1934): 429.

The result has been that passenger trains are again 'hot news,' capable of making the front page, and millions of lines of newspaper and other publicity for the railways have resulted." "Mile-a-minute trains," the article continued, "can no longer be classified as experiments. They have proven their worth as passenger traffic attractions."[36] Another New England streamliner of note was the Flying Yankee of the Boston & Maine and Maine Central Railroads, which ran between Boston and Portland, Maine, and then on to Bangor.

In 1934 the Chicago, Burlington & Quincy Railroad introduced the first of its "Zephyr" streamliners. It and those that followed were built by the E. G. Budd Manufacturing Company of Philadelphia. The company, one of several founded by Edward Budd to supply the auto industry, had mainly sold car bodies to automakers. The firm had perfected their manufacture using steel alloys, which when cold-rolled were not only stronger than carbon steel but could be "pressed into deep-drawn, graceful shapes." Newly developed electric arc welding made fabricating large railroad car bodies possible. The metals used were also "stainless" in that they did not rust, thus reducing maintenance.[37] Indeed, the silver gleam of the unpainted rail cars became iconic, substantially underpinning the nation's idea of what streamlined trains were all about. The Reading Railroad's "Railbus" had been something of a prototype, but Budd engineers went well beyond for the Burlington. They produced lower-slung, lightweight car bodies, the sides and roof of which were made integral to car support. (They were not, in so many words, just boxes with lids.) Car design thus allowed the elimination of the heavy center sills and concrete subfloors of orthodox railway passenger cars without sacrificing safety or stability.[38] And, of course, greatly lightened cars meant lower operating costs. General Motors supplied innovative two-cycle diesel engines. The first train set, placed in service between Kansas City and Lincoln, Nebraska, doubled the revenues of the steam train previously run on the route, with operating costs being slashed from some sixty-four cents to thirty-four cents per train mile. The train set carried a $200,000 price tag. In its first year of operation it earned some $97,000, eventually recouping its entire cost in a mere twenty months.[39] The new engine was also much cleaner than earlier diesels, using, as it did, newly developed distillate or diesel fuel. Thus, cylinders did not foul as readily, and trains did not emit oily smoke.

FIGURE 7.9. Advertising postcard for the Burlington's Denver Zephyr, circa 1935. Authors' collections

The original Burlington train set was put in service between Kansas City and Lincoln, Nebraska, and named the" Pioneer Zephyr." It was also displayed across the nation and then also sent to Chicago's Century of Progress Exposition in 1933. It set a long-standing speed record on its way in from Denver: some 1,015 miles in thirteen hours and five minutes at an average speed of 77.5 miles per hour. It hit a top speed of 112 miles per hour.[40] The Burlington put in operation the nation's largest streamliner fleet, beginning with the Twin City Zephyr (Chicago to Minneapolis–St. Paul) and the Mark Twain Zephyr (St. Louis to Burlington, Iowa, via Hannibal, Missouri), and the Nebraska Zephyr (Chicago to Omaha and Lincoln). Eventually there were nine others. Pictured is the Denver Zephyr (fig. 7.10). The trains were named by Ralph Budd, president of the Burlington, for Zephyrus, "the god of the west wind and symbol of rebirth."[41] The articulated-car train set succumbed almost as quickly as it arrived, although many did continue to operate through World War II.

Illinois Central

Green Diamond

Weight of train:	480,200 lb.
Consist:	5 cars—1 power unit; 1 mail-baggage-express; 1 coach; 1 coach-diner; 1 diner-lounge.
Placed in service:	May 17, 1936
Operated between:	Chicago and St. Louis (1 round trip daily)
Daily mileage:	588.4
Overall scheduled speed:	59.9 m.p.h.
On time percentage	99 per cent

Statistics
(May 17, 1936 to June 30, 1940)

Total train miles:	809,289
Total passengers:	241,909
Total passenger miles:	50,370,426
Average passengers per train:	62.24
Revenue per train mile:	$1.48

FIGURE 7.10. The Denver Zephyr, Circa 1935. Postcard, authors' collection.

Operators of long-distance trains simply needed flexibility in adding and subtracting coaches as traffic ebbed and flowed. Articulated trains were inflexible.

Many self-propelled rail motor cars continued in service, joined by new streamlined versions. As they had been originally intended, they continued to be employed mainly on surviving branch lines, but additionally on short commuter runs. In 1940 the New York, Susquehanna & Western Railroad introduced self-propelled, eighty-passenger coaches manufactured by the American Car & Foundry Company at its Berwick, Pennsylvania, plant. Employed in commuter service, train schedules were carefully coordinated with those of the company's bus subsidiary. Indeed, rail cars and buses were very similarly styled in a modernist manner, painted in the same colors, and given the same logos.

The Illinois Central Railroad purchased American Car & Foundry cars for its "Illini" between Chicago and Champaign, Illinois. The train covered 183 miles, making sixteen stops in roughly four hours at an average speed of forty-six miles per hour. It was intended to provide local service on its main line in order to free up, and thus speed up, its crack long-distance passenger trains connecting Chicago with Memphis and New Orleans.[42] To the south, the Illinois Central adopted a modified version of the cars for the "Miss Lou" which ran between New Orleans and Jackson, Mississippi. With side entrances as well as entrances up front, coaches were designed in what the railroad termed its "Jim Crow" arrangement. Blacks entered at the middle of a car and sat in the rear. Whites entered up front and sat up front. Two sets of restrooms were provided. The two sections were separated by equipment rooms containing air-conditioning and brake compressors as well as a small buffet counter. The IC's new rail cars carried sixty-nine passengers up north but only sixty down south.

LATER STREAMLINERS

The streamlining idea then transferred to standard passenger-train equipment. New passenger cars were built with the light metals and framing techniques. Steam locomotives were dolled up to appear modern with streamlined exteriors. New diesel locomotives were similarly given the streamlined look. Streamlined trains symbolized speed, and in long-distance travel speed clearly counted. Streamlining and speed, the railroads speculated, could be had with such modified standard equipment. For many years travel times on main lines had been decreasing as speeds were increased. Many railroads had introduced "extra-fare" passenger trains with express service between cities. Express service on the Pennsylvania Railroad between New York and Chicago operated on a twenty-four-hour schedule up until May 1932. In July 1933 it was down to twenty-one hours and fifty-two minutes.[43] New equipment helped. On the Pennsylvania's Broadway Limited, it was improved steam power and lighter-weight passenger cars. On the Illinois Central, it was the Green Diamond between Chicago and St. Louis, with new diesel locomotives and new lightweight cars (fig. 7.11).

As the Zephyrs went into service on the Burlington between Chicago and Minneapolis–St. Paul, competing railroads kept pace, but streamlined standard equipment. On the Chicago & Northwestern it was the "400," with a modern veneer on a standard steam locomotive and new lightweight standard cars. On the Milwaukee it was the "Hiawatha" with a new oil-burning steam engine and streamlined cars. The year 1936, reported *Railway Age*, was a year of outstanding traffic developments. "A record-breaking inauguration of streamlined passenger trains, a further widespread speeding up of both passenger and freight train schedules, the air-conditioning of more passenger cars and a marked extension of pick-up and delivery service for merchandise were the major contributions of the railways to the improvement of passenger and freight service," noted *Railway Age*. Fourteen new passenger trains had been inaugurated in the United States during the year, most of them with improved lightweight standard equipment. Included were the Super Chief on the Santa Fe and the City of San Francisco on the Chicago & Northwestern, Union Pacific, and Southern Pacific. The Chicago, Rock Island & Pacific had launched six new "Rocket" streamliners.[44]

An Interstate Commerce Commission study compared lightweight diesel-powered passenger trains with those pulled by steam locomotives. According to *Bus Transportation*, "On a seat-mile basis, diesel wins, showing average total direct cost of \$0.0030 as compared with \$0.0035 for steam. Rail motor cars were higher than either at \$0.0037."[45] But equipment was just one factor in improv-

FIGURE 7.11 . Illinois Central's Green Diamond. "March of the Streamliners," *Railway Age*, n.s., 109 (Nov. 16, 1940): 725.

ing speed and lowering costs. New equipment demanded improved rights-of-
way and, especially, heavier rails if not welded rails. Curves, grades, and grade
crossings limited speed. So also did rough track due to poor repair. Improved
maintenance required new power equipment. "These include," noted a *Railway
Age* reporter, "tie-tamping equipment, bolt tighteners, spike pullers, spike drivers, tie adzers, rail grinders, ballast cleaners, ditching and trenching machines,
draglines, power shovels, welding outfits, rail cranes, locomotive cranes, bridge
derricks, pile drivers and other machines and tools suitable for constructing and
maintaining bridges, laying rail, ballasting, renewing ties, draining the roadbed
and day by day routine track maintenance." Additional turnouts or passing tracks
were necessary for high-speed service. Power-operated, remotely controlled signals and switches were necessary. For diesel-powered trains, new fueling facilities were needed. For faster, more efficient steam locomotives, coaling and water
facilities needed to be relocated farther distances apart.[46]

And new passenger depots were overdue in many cities. A long period of declining passenger revenues had discouraged passenger station remodeling. "The
result has been that in the face of increasing passenger traffic, patrons are still being routed through stations that are obsolete, poorly maintained, generally unat-
tractive and, in some cases wholly unfit for the purpose, to reach trains of the latest
and most attractive designs, which have been highly dramatized in advertising
and other forms of publicity," said *Railway Age*. Needed were stations that, like
the trains, used the latest building materials as well as "the most modern forms
of florescent lighting, with heating and plumbing to correspond."[47] What had the
streamlining of passenger trains wrought? What had the rail motor car wrought?

We think today that streamlined passenger trains of the early twentieth century
came out of aviation. Certainly the streamlining idea did. But boat building also
contributed. And streamlining, it must be remembered, also came to influence
automobile design, so significant did air travel and airplane design become in
the late 1930s. But for railroading it was the rail motor car that first put streamlining forward on the ground. The rail motor car was part and parcel of steam
railroading's adaption to circumstances wrought principally by the coming of
automobiles and motorized trucks and buses. The outright adoption of the internal combustion engine was significant. Early on, it was the gasoline engine

designed to drive bus-like vehicles on rails, with power transmitted mechanically to their wheels. Then it was power transmitted to electrical motors to turn
the wheels. Then came the perfection of the diesel engine. Importantly, it was
automobile manufacturers, their parts suppliers, and their industrial adjuncts
(like car-body manufacturers and tire makers) that actually did much of the
manufacturing when it came to new high-speed streamliners.

The rail motor car eventually succumbed not just to automobiles but to
automobiles and improved highways. They made railroad branch lines obsolete and would eventually make many a railroad main line unprofitable as well.
Automobility was something the railroads could not ignore. And they did not.
They substantially adopted and adapted, with the rail motor car being one of
the most important innovations. The railroads not only took technology from
the automakers but also tried to embrace the values that underlay motor vehicle
popularity, especially that of speed in transport. Unfortunately, the railroads
could not match the motorist's quest for freedom of action that the idealized
"open road" promised. Trains remained fixed to set routes and set timetables.

Chapter 8

RAILROAD BUS SUBSIDIARIES

The motor bus, like the self-propelled rail motor car, provided a means of reducing operating costs on marginal railroad branches. They also presented a means of extending service beyond existing lines, thus to funnel additional customers to passenger trains. Out of railroading's involvement with busing came the nation's largest bus systems—Greyhound and Trailways, companies that rose to dominate intercity bus travel in the 1930s. Along with their trucking interests, railroad bus subsidiaries came close to making several of the nation's largest railroads into integrated transportation companies. Even air-passenger service was for a short period of early experimentation commercially linked to railroading, although not through ownership but through partnership. The rise of the nation's bus industry and the role played by various railroads in its evolution is the story told herein. It begins with the private automobile's undercutting of railroad passenger traffic. It continues with the adoption of motor buses as a means of cutting railroad operating costs, with buses being used first to replace expensive-to-operate steam locomotives and standard passenger car equipment on short branch lines. It ends with the railroads underpinning long-distance bus travel nationwide. The route the railroads traveled was one of sustained opposition to buses as common carriers, but also one of ever-increasing involvement in the industry, even to the point of dominating long-distance bus travel. Busing was an important form of railroad adaptation to the nation's preference for motoring.

DECLINING RAIL PASSENGER TRAFFIC

Ridership on the nation's Class I railroads peaked at some 1.2 billion in 1920 and then began a steady decline to some 700 million in 1930, a decrease of 43 percent.[1] Decline came more from those traveling short distances than from those traveling long distances. It was, of course, the automobile's use in short-haul driving that made the difference, but also it represented the growing importance of busing in Americans' commuting to work and journeying to shop. This was especially true of southern New England, with its high population densities and towns and cities only short distances apart. It was there that railroad busing first took root with vigor. It was there that railroad executives were most receptive to adopting buses as a means of countering losses attributable mainly to the rapidly increasing popularity of motoring. The Interstate Commerce Commission estimated that private autos took between 80 and 90 percent of lost railroad passenger traffic. But buses took most of the rest. Beyond New England, however, the Denver & Rio Grande Western ascribed about one-sixth of its passenger loss to busing, the Southern Pacific one-quarter, and the Louisville & Nashville over one-third.[2]

The editors of *Railway Age* dutifully reported year by year the damage done. The number of automobiles registered in 1924 stood at 15,460,649 and increased annually over several years at a rate of about 20 percent. "The saturation point for automobiles has apparently never been reached," they noted. "On January 1, 1925, there were 2,866,061 miles of highways, of which 470,000 were surfaced roads. . . . the total expenditure for highways in 1924 being $990,683,770. Railway officers do not need to be told that a not inconsiderable portion of their taxes is thus being used to make it easier for people not to use the railroads."[3] Railroad executives began issuing wake-up calls in public speeches and articles published both in railroad trade journals and popular magazines. In 1925 Ralph Budd, president of the Great Northern Railway, emphasized that train ridership across the nation had declined by one-third in just a few years, while at the same time the country's population had increased some 15 percent. The situation, he said, called for the railroads to adopt buses to help cut accruing operating losses: "The convenience of highway travel is responsible for the motor coach, which in its best development affords flexibility as to route, schedules, and service as well as riding comforts comparable with the automobile, and at a cost comparable with railroad fares."[4]

Between 1920 and 1927, noted W. W. Atterbury, president of the Pennsylvania Railroad, the passenger revenues of Class I railroads had declined by 24 percent with an average annual loss of some $40 million. In the same period, auto registrations had increased some 150 percent. For every 6 percent increase in automobiles there had been a 1 percent decrease in passenger revenues. In addition, operating revenues taken away from the railroads by motor trucks had amounted annually to approximately $132 million, and this despite the fact that the railroads were carrying an ever-increasing amount of automobile-related freight. In 1927 alone, the railroads had moved over 757,000 carloads of automobiles, trucks, and parts, which constituted the third largest category of manufactured products carried. When such things as petroleum, factory raw materials, and highway construction materials were factored in, some 3.2 million carloads were "traceable to the manufacture and use of motor cars." Thus his railroad was purchasing "substantial interests" in both truck and bus companies. Regarding busing, the Pennsylvania had just bought the Philadelphia Rapid Transit Company, which itself had just purchased the Peoples Rapid Transit Company. Together the two lines connected New York City, Philadelphia, Atlantic City, Baltimore, and Washington, D.C. Both electric railways had bus subsidiaries that were being integrated into the newly incorporated Pennsylvania General Transit Company.[5]

MOTOR BUSES

The motor bus derived from the motorcar and quite literally so. In 1915 C. J. Ecklund and his brother, two Swedish immigrants working in Minneapolis, produced what might have been the nation's first buses by cutting apart large sedans (such as those manufactured by Pierce Arrow and Packard) and then inserting frame extensions and extra body panels to substantially lengthen them to accommodate twelve passengers. In 1921 the Fageol brothers of Oakland, California, introduced what may have been the first vehicle designed from scratch to be a bus, or what initially came to be called an "auto bus."[6] Numerous companies began manufacturing bus-like vehicles. For example, the Chicago Pneumatic Tool Company introduced in 1911 its Little Giant "commercial car" with a 20-horsepower, internal-combustion, water-cooled engine, solid rubber tires, and double-side chain drive. It could carry ten passengers, "making it particularly desirable for station and hotel use."[7] Pictured in a postcard view is a larger but similar vehicle used to carry sightseers in Denver (fig. 8.1). The card's caption reads: "Seeing The Foothills By Trolley Excursion Automobiles." A handwritten message reads: "Seeing Denver, July 10, 1912."

Motor bus technology evolved rapidly but not without many false starts. The Versare Corporation of Albany, New York, perfected a vehicle that moved "on two four-wheel trucks" and could "seat 44 passengers with standing room for 52 more," *Railway Age* reported n 1925. It was promoted as "a heavy highway car for railroad use." Indeed, it was very much like a rail motor car except it was intended for highway driving. Its drive was gas-electric and, indeed, it ran on "trucks" much like those of railroad passenger cars, except that the wheels were of hard rubber. "Both pairs of trucks are movable as are also, separately from the trucks, the front pair of wheels of each. The result is a highway vehicle of high capacity which, in spite of its size, can still negotiate narrow and crooked streets with ease."[8] The editors of the *Electric Railway Journal* described it as a vehicle of "medium capacity body with semi deluxe appointments, mounted on a low chassis of long wheel-base." It gave "an exceptionally pleasing general appearance," the journal said, adding, "Approaching the private auto in speed, exclusiveness and general appearance and making special appeal for patronage to the man who usually 'drives his own,' this type of bus should fit in exception-ally well in providing fast express service."[9] Standardized auto and truck engines, transmissions, and chasses came to dominate bus manufacture. They were pro-duced by many of the leading auto and truck makers, including White, Mack, Studebaker, Graham, and others. Most bus manufacturers limited themselves to building bus bodies and final bus assembly, with the bus bodies being lowered onto chasses with the engines already installed. Railroad-like vehicles largely disappeared from the nation's highways. Pictured is a 1926 advertisement for the REO Motor Car Company (fig. 8.2). Note the market that it specifically targets. It is headlined, "REO BUSSES, Feeders to the Railroads."

The cost per mile to operate a bus was approximately one-fifth that of oper-ating a train even when the latter comprised only a locomotive and a single car, which had become in the 1920s the norm on most railroad branch lines. The cost ratio of highway bus operation to steam-train operation, Ralph Budd noted, meant that for "the cost of one train in each direction, say morning and evening, a bus could be run every two hours in each direction from 8 a.m. to 4 p.m." And such service better suited the needs of the typical rural community. "Owing to the extensive use of the private automobile there is scarcely enough travel even morning and evening on the average local run to justify a train, much less to justify several trains during the day; but the smaller and less expensive motor bus operating on the highway may pick up sufficient traffic to make it profitable," he argued. Furthermore, there was the advantage of more convenient starting and stopping. Buses could pick up and drop off passengers anywhere along a route. And routes could be changed to meet changing needs. Buses were more flexible.[10]

By 1921 there were some 32,000 motor buses in operation in the United States, most in direct competition with steam- and electric-railroad passenger service. *Railway Age* declared, "The number of bus companies now operating and the number of routes they cover are astounding." For example, in the New England states, some 350 bus companies operated over some five hundred routes. In the three Pacific coast states alone, some 700 bus companies operated over twelve hundred routes. The president of one eastern railroad estimated that fifty-two motor bus routes operated in direct competition to his railroad's passenger lines. "We operate 27 local passenger train routes and one or more bus routes are in

FIGURE 8.1. "Trolley Excursion Automobiles," Denver, Colorado, 1912. Postcard, authors' collection.

FIGURE 8.2. Advertisement for the REO Motor Car Company.
Railway Age, n.s., 80 (June 26, 1926): 38.

competition with 26 of these," he said.[11] It was time for railroad executives to recognize the threat that motor buses and not just automobiles represented— and to recognize as well how they themselves could benefit from bus technology. Never was travel by motor bus as popular as travel by automobile. But for those who could not afford autos or for those who disdained motoring for certain kinds of travel—commuting to work, for example—buses did represent an important travel alternative that the railroads, as well as electric interurban cars and streetcars, could not equal. How might the railroads cash in?

Intercity buses were becoming more comfortable. As one journalist noted, "Most of us have ridden in their deeply cushioned individual seats, reclining at as much ease as in our own cars, looking at the countryside through the large, clear windows or reading at night by the profusion of electric lights. We have stowed our luggage in their handy, roomy baggage compartments. We have stepped into them at our front door or from the porte cochere of our hotel, and then have been transported—swiftly, comfortably, and safely."[12] Of course, the nation's new highways helped. Up through the early 1920s, the motor bus was not recognized as a specialized vehicle fully distinct from any of its predecessors, those being not only the automobile and the streetcar but, technologically, the motor truck as well. Again, buses consisted of a standard truck chassis upon which was mounted some kind of body (usually wood-framed but sheathed in sheet metal), inside of which bench seats were variously arranged with little concern for comfort. But now deluxe sedan-type, low-slung, and fully metal-framed buses with more powerful engines and more dependable transmissions were being adopted. They were being provided with baggage compartments and even with kitchen and restroom facilities. Heating for winter travel and even air cooling for summer travel was becoming standard. And by 1928 even sleeper coaches with berths (not unlike those of railroad Pullman cars) were being placed in service.[13] Safer and more comfortable bus suspension augmented the nation's smoother roads. The bus had come into its own.

For many railroads rate-cutting was the initial response to increased bus competition. Indeed, new "excursion" and other below-cost passenger rates continued for some lines well into the 1930s, as evidenced by trade-journal headlines such as "Fare Cuts Presage New Bus-Rail War in Texas."[14] Bus companies, by and large, only responded by cutting rates themselves. Lowered rates, one bus authority maintained, were "wholly unjustifiable but necessary to meet rates initiated by the railroads." He continued, "The bus people feel that the new rates are intended to put the bus lines out of business and reclaim for the railroads the patronage lost to competitors." Such was not about to happen. Buses had only taken about 5 percent of rail passenger traffic, he said. The chief competitor of the passenger train, and of the motor bus as well, was the "individually owned automobile."[15] Lowered fares changed little. Slightly more ridership was attracted both to train and bus travel, but percentages of increase were for each mode very much the same. For the railroads, with their higher operating costs, rate wars only increased their operating deficits.[16] With the nation in deep economic depression, other initiatives were called for. What the private automobile offered (and, increasingly, the motor bus as well), the railroads needed to offer: convenience, service, comfortable equipment, frequency of schedule, and, above all, speed. The rise of the passenger streamliner would offer one important railroad innovation. Increased use of motor buses would offer another.

Thus it was that the nation's railroads turned to establishing their own motor bus services. And they did so very rapidly. By 1930 Class I railroads in the United States had some $40 million invested in motor vehicle operations, led by the Southern Pacific, which alone had invested some $12 million. Runners-up included the New York, New Haven & Hartford and the Pennsylvania Railroad, each with some $6 million invested, the Union Pacific with $4 million, the Chicago & Northwestern with $2 million, and the Missouri Pacific with $1.7 million. Over twenty-eight hundred buses were being operated both by steam railroads and by electric interurban lines over some fifty-nine thousand miles of route. In 1930 they carried some 34 million passengers, generating some $14 million in revenue.[17] Accelerated involvement in busing was undoubtedly based on the railroads' desire to reduce operating costs by replacing trains with motor coaches and, as well, the desire to funnel passengers onto surviving passenger trains by using new bus routes as feeder lines. But the rapid growth of railroad busing reflected several other developments. One was the improved motor coach. Another was the extensive, ongoing highway-building program. Still another was the threat of increased state regulation over bus transport and even the coming of federal regulation. The latter was something the railroads themselves had long advocated. As the Great Depression deepened, railroad revenues plummeted through loss both of passenger and freight traffic. Staggering under debilitating ICC regulation, railroad executives found it impossible to respond rapidly, whether through dropping and adding rail service, changing rail passenger and freight rates, or otherwise innovating to solve problems. Thus Congress at long last began to deliberate on the role that highway motor-carrier advantages played in railroading's demise. Should interstate busing and trucking be placed under the purview of the Interstate Commerce Commission? If so, should not the railroads move quickly to establish a commanding role in busing before such regulation came to pass?

In 1929 the railroads did benefit from one important development. The ICC ruled that it could not, at least under existing law, forbid a railroad from creating a bus operation. Nor could it regulate such an operation once it was established.

It was the result of a petition to the ICC challenging railroad involvement in the bus industry. Columbia Stages operated forty buses between Portland, Astoria, and Seaside in the state of Oregon. When the Spokane, Portland & Seattle Railroad incorporated its bus subsidiary and then extended bus service to Astoria over a distance well beyond is rail operation at Portland, the bus line filed its petition, claiming that the railroad's bus service violated the Transportation Act of 1920. That act required railroads that wished to extend line service to obtain from the ICC a "certificate of convenience and necessity." The railroad had not. The bus company also claimed that the railroad's articles of incorporation did not permit it to organize and operate a bus service.[18] Columbia's petition was rejected on both counts.

MOTOR BUS REGULATION

Railroads built and repaired their own rights-of-way, which then were taxed by local governments. Motor buses and trucks did not. "The railways operate on highways that they themselves construct and maintain while buses and trucks operate over highways which the public constructs and maintains." So went the argument that the railroads repeated over and over. Furthermore, bus and truck lines were allowed in most parts of the country, it was argued, to "engage in commercial competition without any of the strict regulations of rates and services to which the railways were subjected."[19] But by 1926 thirty-eight states had indeed adopted some kind of motor-carrier oversight, something the U.S. Supreme Court had validated. Much of it was directed at vehicle safety. Restrictions might cover vehicle weight, length, width, and height, as well as the nature of bus braking, venting exhaust, interior air circulation, lighting, and fuel storage. The accompanying set of diagrams illustrates, by way of example, how bus body regulations varied across certain states, much to the chagrin of bus manufacturers who had to confront a vast array of differing rules in marketing buses from state to state (fig. 8.3).

Having evolved out of auto-jitney service, many early bus lines not only operated unsafe vehicles, but did so on constantly shifting routes with erratic schedules and ever-changing rates, their purpose being to skim the cream off of street-car, interurban, and then rail traffic. Consequently, most states came to license bus operators, restricting them not only to operating government-inspected vehicles but also to operating on set routes with set schedules at set rates. Additionally, most state regulating bodies, usually the state utility commission, came to require certifications of necessity, with applicants for bus licenses thus required to prove public benefit. Evidence was needed that new bus service would not hinder or undermine an already established service, be it by rail or by bus. The editors of *Railway Age* approved: "Just as proper restrictions are desirable to prevent wasteful and uneconomical competition between the automobile and the rail carrier, it is even more essential between motor carriers themselves, if strong dependable, financially responsible operators are to be maintained in the field for the benefit of the whole populace." Otherwise, so-called wildcatters would "disrupt and destroy the legitimate business of a lawful operator."[20] But ten states did not regulate bus operations in 1928. And many of the states that regulated did so lackadaisically; oversight of bus inspections, for example, was left largely to bus operators themselves.

Despite the fact that they were now involved in busing and would need to abide ICC involvement, the railroads continued the drumbeat to bring the bus industry under federal mandate. The declaration printed in a 1932 issue of *Railway Age* entitled "Unfair Highway Competition" was typical:

> Railways have suffered staggering losses in business to their competition on the highways. The complete regulation to which the railways are subject renders them incapable of that quick action in meeting new conditions which they should take to defend themselves against the raids of their competitors. Motor vehicle operators, on the other hand, are relatively unregulated, free to snipe at the more lucrative kinds of traffic, free to make any rates they see fit, free to discriminate between shippers and communities, and free to begin or cease operations with no thought of the obligations of a public carrier to the public. The railroads pay their own way and contribute heavily to the support of government. The highway carriers secure their roadway—the public highway—in return for a low rental charge and are thus relieved of a substantial part of their true costs of operation. This enables them to offer what appears to be a cheap form of transportation, but which instead is simply a form of transportation which saddles part of its operating costs upon the taxpayer.[21]

The statement was accompanied by a cartoon.

The bus industry, perhaps rightfully so, charged railroad interests with exaggerating railroad- and bus-operating inequities. Clearly, bus regulation, at least at the state level, had come to the fore. And taxation was not all that skewed to railroading's disadvantage, it was argued. The charge that that taxes paid by railroads

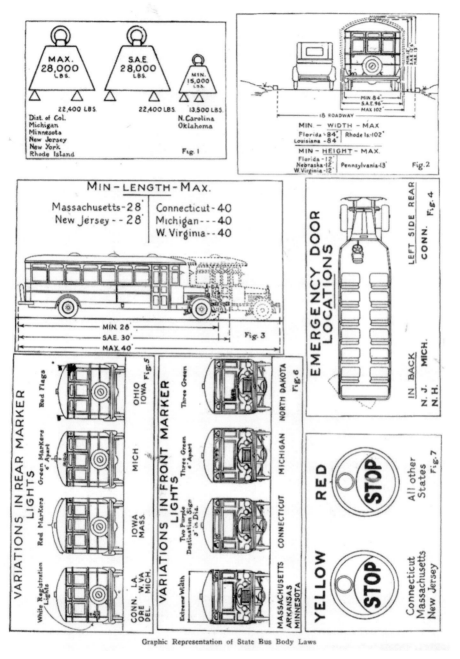

FIGURE 8.3. Graphic depiction of varying state laws regulating bus bodies. Merrill C. Horine, "Buses and Laws Affecting Them," *Railway Age*, n.s., 71 (Sept. 25, 1926): 602.

were supporting highway construction appeared to be overstated. As Frank Fageol, president of the Twin Coach Corporation of Kent, Ohio, noted in a guest article in *Railway Age*: "It is rather a surprise to most people, but, nevertheless an interesting fact, that of the $1,465,000,000 spent on highways in 1927, only $35,000,000 was paid by the railroads, or 2.4 per cent; in other words, less than 10 percent of the total taxes paid by the railroads are used for highway purposes. In the past, I think that many of the rank and file of the railroads have been prejudiced and prevented from trying to get the proper grasp of motor coach development by their erroneous belief that the common carrier motor vehicle was not paying its way."[22] Indeed, the ICC rebuked the nation's Class I railroads for expending nearly $85 million for "propaganda against highway transportation" between 1930 and 1933.[23]

While generally opposing the bus industry and its perceived privileges, railroad interests organized to promote their own busing interests. In 1926 the Railroad Motor Transport Division of the American Railway Association was created. Meeting annually in their own convention were representatives of the nation's various railroads having bus operations. At the St. Louis meeting in 1929, discussion centered on a variety of concerns, among them regulation of motor carriers, highway taxes, methods for computing rates for bus depreciation, unit maintenance methods and costs, and the handling of the U.S. Mail. In 1932 officers included the presidents of several railroads, the president of the National Association of Motor Bus Operators, a vice president of General Motors, the president of the Texas Company (refiners of Texaco products), and a member of the executive committee of the American Automobile Association.[24] *Railway Age,* very much still the industry's leading trade journal, launched a "Motor Transport Section" in each monthly issue.

RAILROAD BUS OPERATIONS

Railroads conducted bus operations in a variety of ways. First, and probably most common, a subsidiary motor bus company was incorporated as a totally separate financial entity, the railroad being a holding company with control vested through stock ownership. Second, through an operating division fully within a railroad's chain of command, the railroad in essence operated buses itself. The third mode of operation was through an affiliated company, the ownership of which might be shared by several railroads. The fourth method was through contracting with fully independent bus companies to provide bus service. One reason for favoring both independent bus subsidiaries and contracting with independents was the labor situation. Railroad employees were governed by the 1907 Hours of Service Act (overseen by the ICC), which limited the hours a person engaged in railroad work could remain on duty. Railroad employees also enjoyed high wages and favorable working conditions, thanks to their strong labor organizations. Bus company employees were notoriously underpaid, benefiting from neither ICC oversight nor union organization. And their work days were notoriously long.

Railroads adopted buses for a combination of reasons. Foremost, of course, was cutting costs on branch-line operations. The editors of *Bus Transportation* reported, "Practically all the railroads using buses have effected economies by substituting them for some of their unprofitable branch line trains. In most instances the yearly operating loss of the former rail line has been considerably reduced by this substitution. In some cases the buses have actually turned a profit." Buses were also used, of course, to eliminate train stops on main lines: "By using buses to pick up passengers at poorly patronized stations and carry them to the nearest depot not only is a saving made in actual operating expenses, but faster train schedules are made possible." The editors continued, "In sections of heavy seasonal passenger traffic higher train speeds and more frequent headways can be rendered by the elimination of local trains and in their place using buses over highways." The use of buses in establishing feeder service was important: "Buses have been used to feed express trains covering a radius of from 15 to 20 miles beyond the point where the train schedules terminate." Buses were used on new routes totally independent of train service: "In some ways [the] most important development came when the railroads definitely entered the fields of long haul bus transportation as competitor for passenger traffic quite apart from their rail lines." Significant, too, was the use of tour buses to divert rail passengers into slower paced sightseeing: "Buses are used for sightseeing purposes, while trains are utilized for quick travel from one region to another."[25] So also were buses used in emergencies to carry passengers otherwise stranded when rail service was disrupted. In many instances, however, bus services were, as we have seen with the interurbans, started mainly to prevent an independent bus operator from initiating a competing service, the railroad choosing to compete with itself rather than allow another corporate entity to do so. Buses, of course, were used to carry railroad employees to work locations.

In 1929 thirty-three Class I Railroads variously operated some 3,100 motor buses over some 66,000 miles of route, 38,000 miles of which were *intra*state and 28,000 *inter*state. In 1930, eight railroads operated 100 or more buses: the Southern Pacific (648), the New Haven (638), the Pennsylvania (430), the Great Northern (218), the Missouri Pacific (192), the Union Pacific and the Chicago & North Western together (188 apiece), the Texas and New Orleans (a Southern Pacific subsidiary, with 147).[26] Railroad busing varied substantially not only from railroad to railroad but from one region to another, with the American Northeast and especially New England leading the way.

Motor buses made sense. F. J. Scarr, owner of a bus company in New York, wrote, "The motor coach is designed to handle small groups; it is highly flexible as to time of movement and route of travel, is not dependent upon rigid schedules, and approaches the convenience of individual service. The steam train is designed to handle large groups of passengers; generally operates at higher speeds, and is confined to fixed schedules and routes." There was another important bus characteristic to note: "The motor coach, moving through the main streets of cities and towns, rather than the 'back yards,' and traversing generally more pleasant routes, free from smoke, cinders and noise, adds attractiveness as well as convenience to travel, thereby enhancing the intangible, but certain, desire to 'ride on rubber.'" The bus, Scarr argued, was not really a competitor. "The motor coach, advancing into the field of organized highway transportation, has drawn its traffic largely from the private automobile, and as traffic congestion and parking regulations become more severe and a more general knowledge is attained by the private automobile owner of the cost of supplying himself with this convenience the motor coach will draw its patronage more and more from the private automobile rider."[27] That, of course, would prove to be wishful thinking.

NEW ENGLAND

The New York, New Haven & Hartford was different from most other American railroads in that passenger traffic generated a substantially higher proportion of its revenues, its lines being concentrated between Boston and New York, an area of very high population density. In 1925 the relation of freight to passenger income was approximately 55 to 45 percent, and nearly one of every eleven railroad passengers in the United States was carried on the New Haven.[28] The railroad was the result of many short lines being consolidated, perhaps too many lines having been built in the first place, with much duplication resulting in the region. Elimination of branch-line service, and then of branch lines themselves, had become a high priority. Study revealed that, on average, passenger trains on the New Haven carried about 150 people. The cost per mile of operating rail service stood at $1.23 per mile. Yet many branch line trains had fewer than twenty-five passengers, and earned below thirty cents per mile.[29] A subsidiary, the New England Transportation Company, was organized to operate buses on pubic highways, thus to eliminate unprofitable passenger trains, extend service beyond rail lines, and eliminate local stops on main lines. Employees displaced in the shutting down of train service were given preference for jobs servicing and driving the buses of the new company.

Operations began in 1925 with six Mack buses and twenty more on order from Yellow Coach (owned by General Motors), sixteen more from Pierce-Arrow, and fifteen more from the Fageol Company, which had just been bought by a maker of rail cars, the American Car and Foundry Company.[30] Although it would eventually be recognized that having fewer kinds of buses meant real savings in maintenance and parts replacement, buying different kinds of buses initially resulted from recognition that different routes had different needs, including varying bus sizes. In 1927, however, the company did adopt a standard bus body, which would be attached to a chassis manufactured by different truck manufacturers—specifically, Mack, Yellow, White, and International Harvester. The company very quickly established an express route between New York and Boston in direct competition with its principal passenger trains. The 240-mile trip varied time-wise from eleven hours and twenty-five minutes to eleven hours and forty minutes, the differences being a function of time of day and day of week. The buses employed on the route were "parlor observation coaches" assembled in the J. G. Brill Motors Company plant at Philadelphia, a manufacturer of streetcars and electric interurban cars, which had also been recently purchased by American Car & Foundry. By 1928 the New England Transportation Company was operating over some fifteen hundred miles of highway (fig. 8.4). The company continued to expand its routes by petitioning to establish new ones, but also by buying out independent bus operators. In 1939 the company was also operating several subsidiaries of its own: Victoria Coach Line, Berkshire Motor Coach Lines, Providence-Hartford-Norwich Lines, and County Transportation Company, the latter originally a subsidiary of an interurban line, the defunct New York & Stamford Railway Company.[31]

The Boston & Maine Railroad centered on Boston with lines radiating west, northwest and north as well as "down east" to Portland, Maine, where the Maine

Central, its rail subsidiary, extended its reach east to the New Brunswick border as well as north up into New Hampshire. The Boston & Maine faced the same branch-line problems that plagued the New Haven. When asked whether the railroad should be allowed to adopt busing, the chair of New Hampshire's Public Service Commission answered: "Shall the Boston & Maine Railroad stick to its rails and see its position forsake its trains for this modern and popular mode of travel and do nothing to stop it?" The railroad's passenger revenues had declined by some $3.2 million dollars in just three years, 1923–1925. It was indeed time for coordinated rail-bus service, he thought.[32] In the period 1921–1925, when passengers carried on all Class I railroads in the United States declined some 14 percent, the Boston & Maine's traffic had declined 23 percent.[33] Thus a bus subsidiary, the Boston & Maine Transportation Company, became a most welcomed addition. Placed in operation were sixty-one motor buses on twenty-four routes, eleven of which involved train replacements. Whereas it cost the railroad from $1.17 to $2.02 per mile to operate steam trains, it cost the bus subsidiary twenty-six to thirty cents per mile to operate buses.[34] In 1929 B&M Transportation operated

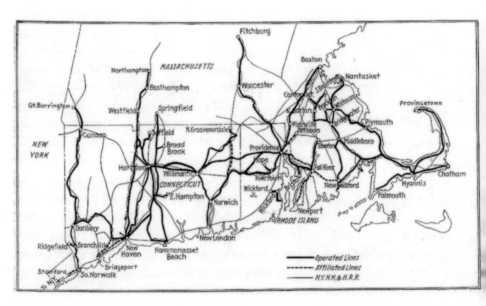

FIGURE 8.4. Routes of the New England Transportation Company. "N.E.T. Systematizes Operations," *Railway Age*, n.s., 84 (Feb. 25, 1928): 493.

ninety-nine buses over thirty routes of some twelve hundred miles.[35] Over the bus company's first four years of operation, its discontinuance of branch-line train service reduced operating expenses for the railroad by some $1.4 million. The aggregate cost of running buses was $916,000.[36] The Maine Central, for its part, first introduced buses mainly to attract summer tourist travel. Accordingly, its subsidiary, the Samoset Company, which owned several resort hotels, had placed ten buses in operation in 1925.[37] The Bangor & Aroostook, the Central of Vermont, and the Rutland were the other New England railroads operating buses.

ELSEWHERE EAST OF THE MISSISSIPPI

The Pennsylvania was the nation's largest railroad. Indeed, it called itself the "standard railroad of the world." Its principal line connected Philadelphia with Pittsburgh, where two main lines with numerous interconnecting branches went on to Chicago and St. Louis, respectively. But another main stem connected New York City with Philadelphia, Baltimore, and Washington, D.C. In 1930 its bus subsidiary, the Pennsylvania General Transit Company, not only operated buses but was also operating a dozen bus subsidiaries of its own. It also controlled the People's Rapid Transit Company, which ran buses from Philadelphia to New York City, Atlantic City, Baltimore, and Washington, D.C. Pennsylvania General's 429 buses were carrying some 3 million passengers annually. Why did people ride its buses? *Railway Age* offered various reasons: (1) improved highways, (2) economy (motor coach fares were about 40 percent lower than rail fares), (3) convenience, (4) passing scenery ("The passenger usually gets a better scenic view of the country than from a train. At any rate he views it from a different and less accustomed viewpoint."), (5) flexibility, (6) door-to-door, center-to-center service, and (7) what was termed the "psychological principal of imitation" ("The widespread use of private motors has created a general desire to 'ride on rubber.'").[38] So also had the Pennsylvania Railroad inaugurated, along with the Santa Fe Railroad, a forty-eight-hour New York-to-Los Angeles passenger service involving both train and air travel. For daylight stretches, travelers flew on planes provided by the Transcontinental Air Transport Company (TAT, which became Trans World Airlines, or TWA).[39] Pundits labeled TAT the "take a train airline." But it was the Pennsylvania General Transit that was the principal story.

Almost every Class I railroad in the American Northeast eventually ran buses, but two are especially noteworthy as pioneers: the Reading Railroad and the Baltimore & Ohio. The main lines of the Reading radiated north, northwest, and west out of Philadelphia with a very high density of branch lines interconnecting cities across southeastern Pennsylvania. Its trains accessed Jersey City over the tracks of the Central Railroad of New Jersey, actually a Reading subsidiary through much of the twentieth century but one that did enjoy periods of independence, a function of repeated Reading bankruptcies. From Jersey City railroad passengers were taken across the Hudson to Manhattan by ferry. In 1930 the Reading-Jersey Central Bus Company was carrying some 1.2 million passengers in seventy-seven motor coaches over 829 miles of route. Buses were costing some $167,000 annually to operate with savings of some $409,000 a year over what the rail operations had cost.[40] The Baltimore & Ohio connected Philadelphia, Baltimore, and Washington, D.C., on the east with both Chicago and St. Louis on the west. The United States Railroad Administration controlled America's railroads during World War I, had extended B&O passenger service beyond Philadelphia to Jersey City over the Reading and Jersey Central, with passengers to Manhattan again being ferried across the Hudson River. In 1927 the B&O arranged for passengers to transfer to buses, which in turn, after having been ferried across the Hudson, continued on to various locations in New York City, including the Grand Central and Pennsylvania Stations and several of the city's principal hotels. Eventually the B&O opened several Manhattan bus stations to replace the hotels. But rather than establish its own bus subsidiary, the railroad contracted with New York City's largest bus operator, the Fifth Avenue Coach Line, to provide the vehicles, with a fleet of coaches being painted in the distinctive blue and white trim of the B&O passenger trains (fig. 8.5).[41]

Railroads across the American South and in the Middle West also launched bus services but certainly not on the scale of the northeastern railroads. Those operating buses in the South included the Central of Georgia; Gulf, Mobile & Ohio; Illinois Central; Norfolk & Western; Norfolk Southern; Richmond, Fredericksburg & Potomac; and the Southern. Not every service was successful. The Nashville, Chattanooga & St. Louis (owned by the Louisville & Nashville) began a small three-bus operation in Tennessee, connecting its railroad station at Cowan with the town of Tracey twenty miles away. The occasion was the paving of the Dixie Highway north from Chattanooga. The service was abandoned after four years. Steady losses had accrued, the improved highway having encouraged use of private automobiles rather than combined rail and bus travel. *Railway Age* editorialized: "The motor coach is not a cure-all, and it cannot in itself completely

FIGURE 8.5. Advertisement for the Baltimore & Ohio Railroad. *A Guide to and from New York* ((Baltimore): Baltimore & Ohio Railroad, c. 1935): front cover.

solve the problem of dwindling traffic on branch lines. The motor coach can be operated with substantially smaller expense than a passenger train, and it can, therefore, effect savings in operating expenses for the railways. But it offers no guarantee that it will stop the decline of traffic. . . . too much should not be expected from the motor coach."[42]

Railroads operating east of the Mississippi River in the Midwest included the Akron, Canton & Youngstown and both the Cincinnati Northern and the Cleveland, Cincinnati, Chicago & St. Louis (two New York Central subsidiaries). So also did the Illinois Central and the Wheeling & Lake Erie. One railroad merits emphasis. The Chicago & Alton's main line connected Chicago and St. Louis with another line splitting off to Kansas City. In 1926 the railroad purchased two eight-wheel highway coaches from the Versare Corporation, putting them in operation between St. Louis and Jacksonville, Illinois. Thus were the first (and two of the very few) gas-electric buses in the country placed in operation.[43] The Chicago & Alton was noted as a pioneer. It had been the first railroad in the nation to operate chair cars, sleeping cars, and dining cars. The first sleeping car, named the "Pioneer," was the work of George Pullman, who for years contracted with the railroad to build his Pullman cars at its Bloomington, Illinois, shops. The Chicago & Alton's pioneering in bus transportation did not prove overly successful despite the company's purchase of other bus lines across Illinois and Missouri, thus to create a long-haul bus system. Midwestern railroads, like those of the South, suffered fewer of the branch-line woes that plagued northeastern railroads. The rail motor car, rather than the highway motor bus, thus remained the preferred means of replacing unprofitable steam trains. And, of course, many of the Midwest's electric interurbans were still operating in the early 1930s.

THE TRANS-MISSISSIPPI WEST

Numerous railroads operated buses west of the Mississippi River across the Great Plains and into the Rocky Mountains: the Atchison, Topeka & Santa Fe; Chicago, Burlington & Quincy (the Burlington); Chicago, Rock Island & Pacific (the Rock Island); Denver, and Rio Grande Western; Great Northern; Missouri-Kansas-Texas (the Katy); Missouri Pacific, Nevada Northern; and Northern Pacific. Also included were the St. Louis, and San Francisco (the Frisco); the St. Louis Southwestern (the "Cotton Belt," which was a Southern Pacific subsidiary); the Southern Pacific itself; the Spokane, Portland & Seattle (owned jointly by the Great Northern and the Northern Pacific); and the Union Pacific (paired in many of its operations with the Chicago & North Western).[44] *Railway Age* carried an announcement as early as its March 10, 1899, edition: "C. A. Higgins . . . of the Atchison, Topeka & Santa Fe informs us that his road is considering the feasibility of putting on a line of automobile carriages to be used between Flagstaff, Ariz. and the Grand Canyon of the Colorado River. One of these carriages is now being constructed in Chicago, and if a test . . . proves satisfactory, other vehicles will be built . . . giving the Santa Fe a more commodious, comfortable and rapid stage transportation to the Grand Canyon than has ever been enjoyed before."[45] This was the first mention of a motor bus in *Railway Age*. But it was not until 1926 that the Santa Fe actually launched a real bus operation, and that was under the auspices of its Fred Harvey subsidiary, which operated Santa Fe passenger-train dining cars and restaurants in railroad stations along its passenger routes.

The Santa Fe was the only line to directly connect Chicago with Los Angeles and San Francisco, its main line running through Kansas City. A secondary line ran south from Kansas City to the Texas Gulf Coast. Early buses were twelve-passenger parlor coaches dubbed "Harvey Cars." Specially designed bus bodies were placed on chasses manufactured by both the White and Yellow truck making companies. Also operated were seven-passenger Packard and Cadillac sedans. Passengers on Santa Fe passenger trains could interrupt cross-country travel to access the Grand Canyon in Arizona. But the most celebrated of the Santa Fe bus services was the so called "Indian Detour" of New Mexico, which featured visits to Native American pueblos in and around the city of Santa Fe.[46] The railroad expanded its bus operations by organizing the Santa Fe Transportation Company, which not only took over the Fred Harvey operation but bought in 1933 the Southern Kansas Stage Lines, which operated not only in that state but in Oklahoma and Texas as well.[47] New bus routes were successfully petitioned in California, where beginning in 1938 buses were used to bridge a gap in its rail network, which had up until then precluded passenger service between Los Angeles and San Francisco.

The Northern Securities Company was formed in 1901 by E. H. Harriman, James J. Hill, J. P. Morgan, J. D. Rockefeller, and others as a means of consolidating various railroads. They included Hill's Great Northern, Northern Pacific, and Chicago, Burlington & Quincy, along with Harriman's Union Pacific, Southern Pacific, and Illinois Central. Northern Securities was sued in 1902 under the Sherman Antitrust Act of 1890 and consequently broken up into its constituent parts, an initiative of the Theodore Roosevelt administration. But the Interstate Commerce Commission had determined that it did not have jurisdiction over the merger of bus lines, even those owned by the railroads.

The Northland Transportation Company, a wholly owned bus subsidiary of the Great Northern, consolidated a host of bus operations in northern Minnesota, laying a base for the rise of the Greyhound bus system. So also did the Burlington Railroad create a bus subsidiary, the Burlington Transportation Company. That service started in 1929 over a sixty-six-mile route between Omaha and Lincoln, Nebraska. The company quickly extended west to Grand Island, Nebraska, and then east to Clarinda, Iowa. In two years it was operating over one thousand route miles using thirty-eight buses.[48] But it was not until 1934 that the company fully embraced long-haul busing, establishing an express route between Chicago and Omaha. With the purchase of the Columbia-Pacific Night Coach Company, the firm then launched a seventy-one-hour Omaha–to–Los Angeles bus service via Salt Lake City.[49] The Burlington Transportation Company would play a leading role in launching the Trailways bus system.

In 1930 the Missouri Pacific's bus subsidiary was perhaps the fastest-growing bus company in the nation. In just four years it had expanded from 63 buses to a 185, and from 260 route miles to 4,000. Buses at first replaced local train service, but then morphed into long-haul operations. The company operated repair facilities at Kansas City, Popular Bluff, St. Louis in Missouri; at Walnut Ridge and McGehee in Arkansas; at Ferriday in Louisiana; and at Houston and Brownsville in Texas. These locations more or less outlined the company's geographical reach. "In the interests of economical operation, the company has erected large gasoline storage tanks at central locations in the various territories in which it operates, and purchases its gasoline in tank car lots at a considerable saving. Lubricants, likewise, are purchased in carload lots," reported *Railway Age*. "At points where the rail stations are not located convenient to the center of towns, or adjacent to the highway, it is, of course, necessary to establish other stations exclusively for motor coach passengers. But wherever possible, the railway stations and the station ticket agents serve both motor coach and rail line passengers."[50]

The Union Pacific Railroad's entry into busing was complex. Through its railroad subsidiary, the Oregon-Washington Railroad and Navigation Company, bus service was established between Portland and Pendleton, Oregon, along the south bank of the Columbia River on the famed Columbia River Highway, still one of the nation's most scenic drives. A bus subsidiary called Union Pacific Stages was then organized to expand eastward from Oregon. It was in Nebraska, however, that Union Pacific's busing operations fully matured. In 1929 independent bus operators (Interstate Transit Lines, Cornhusker Stage Lines, and Queen City Coach Lines) were purchased, a direct response to the bus initiative launched by the Chicago, Burlington & Quincy. Interstate Transit had operated forty-five motor coaches on routes radiating out from Omaha, including service northeast to Fairmont, Minnesota, and south to Kansas City. Cornhusker's twenty-nine coaches had served routes west of Omaha and Lincoln, largely in Nebraska. Queen City's six buses served short routes focused only on Lincoln. Public utility commissions across seven western states (California, Colorado, Kansas, Idaho, Nevada, Utah, and Wyoming) were petitioned for new routes not only to connect with Portland but to reach San Francisco and Los Angeles as well.[51] The Union Pacific's partner in transcontinental rail passenger service, the Chicago & North Western, then entered the picture. Established was Chicago & North Western Stages to connect Omaha with Chicago. However, a new company was also organized under the "Interstate" name to run an express bus service from Chicago through to San Francisco on a route closely paralleling the two railroad's main lines. Continued purchases expanded the Union Pacific's bus system: the Bee Hive Stages (Salt Lake City to Yellowstone National Park), the Gem State Company (focused on Boise, Idaho), and the Utah Parks Company (connecting Cedar City with Zion and Bryce National Parks in Utah).[52]

THE PACIFIC COAST

Most of the railroads operating on the Pacific Coast adopted buses, including the Santa Fe; the Chicago, Milwaukee, St. Paul & Pacific (the Milwaukee); the Northwestern Pacific; the Southern Pacific; the Spokane, Portland & Seattle (a subsidiary of the Great Northern); and Union Pacific. The Southern Pacific Railroad's initial involvement in busing was its California Motor Transit Company subsidiary. In 1926 it operated over some fourteen hundred miles of highway, most of it paralleling main-line tracks from San Francisco south to Los Angeles and from Oakland west to Sacramento. Service was not only coordinated with steam passenger trains, but in the Los Angeles area with the electric cars of the Pacific Electric, another Southern Pacific subsidiary. Railroad stations served as bus stations, many where bus layovers were scheduled with food service. Station ticket agents accommodated both train and bus passengers. Through the late 1920s and into the early 1930s, some 1.2 million bus passengers were being carried annually. Opened or under construction in the larger cities were specialized bus terminals. "While the bus fares are slightly lower than those by rail," *Railway Age* reported, "apparently this does not have a decided influence on the patronage, but the governing reasons for the use of this form of transportation seem to be the frequency and convenience of the service. This is especially true of local runs. The busses . . . will stop any place along the line to pick up and discharge passengers."[53] But California Motor Transit was not the railroad's only initiative. The Southern Pacific bought up other bus companies and also petitioned for new routes, establishing a myriad of bus subsidiaries in the process.

In 1929, three holding companies were created. Pacific Transportation Securities, Inc., and the Pickwick-Greyhound Corporation controlled lines from Portland, Oregon, south and west to New Orleans and also to Kansas City. East of Kansas City, lines were controlled by the Motor Transportation Management Company. Ownership was quite convoluted. Control of Pacific Transportation Securities was held jointly by the Southern Pacific Railroad, the Pickwick Corporation, and the Motor Transportation Management Company, with each having a one-third interest. The Pickwick Greyhound Corporation, the Southern Pacific's leading bus operation, was owned by the Pickwick Corporation and Motor Transportation Management.[54] But matters would become even more convoluted. As *Railway Age* reported, "Nominal control of the Pacific Transportation Securities Company, operating lines exclusively in Southern Pacific territory, is expected to be allocated to the Southern Pacific. . . . The lines included in this grouping are those of the Southern Pacific Motor Transport[ation] Company, the Pickwick Stage Lines, Inc., the Pickwick Stages of Arizona, the Yelloway-Pioneer Stages, the California Transit Company, the Oregon Stages, the Auto Transit Company and the Sierra Nevada Stages. For the present at least these companies will continue to operate under their own names." Some $30 million dollars in capital investment was involved. In aggregate, the various companies operated some one thousand buses over approximately thirty-five thousand miles of highway. An estimated 10 million passengers were being carried annually.[55]

Pickwick Stages, although not originally founded by the Southern Pacific, would prove central along with the other Southern Pacific properties in the establishment of Greyhound. Pickwick, founded in 1912, began transporting sightseers from San Diego to nearby Oceanside, California, with service quickly added both to El Centro in the Imperial Valley to the east and Los Angeles to the north.[56] The line picked up and dropped off passengers under the marquee of the Pickwick Theater in San Diego—thus the bus company's name. In 1922 it expanded north from Los Angeles, offering a seventeen-hour, 455-mile connection to San Francisco. By 1926 service had been extended north from San Francisco to Seattle by way of Portland, and west from Los Angeles to El Paso by way of Phoenix and Tucson.[57] The company was carrying nearly 1.5 million passengers in three hundred buses over five thousand miles of highway and was thought at the time to be the nation's largest bus operation.[58] It was Pickwick Stages that introduced night service between Los Angeles and San Francisco with luxurious buses of its own design and manufacture (fig. 8.6). "The sleeper coach, radically different in design and appearance than anything before built," *Railway Age* noted, "was constructed entirely, including the engine, in the Los Angeles factory of the Pickwick Lines. The stage, designated as a 'Nite Coach,'

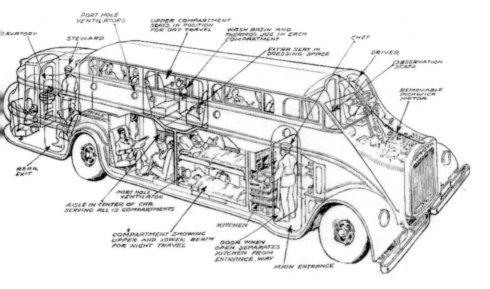

FIGURE 8.6. Diagram of Pickwick Stage's "Nite Coach." "World's First Motor Stage Sleeper Completed," *Highway Transportation,* 18 (Sept. 1928): 5.

consists of two decks comprising thirteen compartments, in each of which two persons may sit or sleep. In addition to the sleeping accommodations, the coach has a complete dining service and lavatory; and there is running water in each compartment." Staggered upper and lower compartments opened onto a center aisle, a single step up or a single step down giving access to each in turn. Heavy sliding curtains gave privacy.[59]

BUS LINE CONSOLIDATION

Railroads, through their bus subsidiaries, greatly influenced the bus industry's coming of age, especially through the great merger period of the 1930s. Railroad-controlled bus companies led in forming Greyhound.[60] A few years later, other railroad bus subsidiaries played leading roles in forming Trailways. Historians place Greyhound's roots in the Mesabi Iron Range. In 1914 a Swedish immigrant, Carl Eric Wickman, established a company to transport iron miners to and from work around Hibbing, Minnesota, in a fleet of motorcars. A year later he partnered with Ralph Bogan to establish Mesaba Transportation Company, connecting Hibbing with Duluth. In 1924 Mesaba merged with the Superior-White Lines of nearby Superior, Wisconsin, to form Northland Transportation Company, with Wickham assuming the presidency. It was Superior-White that first used the Greyhound moniker. The next year an 80 percent stock interest in Northland Transportation was bought by the Great Northern Railroad; Wickham remained as president. Then in 1926 Wickham formed the Motor Transit Corporation, with Great Northern selling back its stock to the company in 1929, at which time Northland become Northland-Greyhound. Then the Great Northern bought 30 percent of Northland-Greyhound's stock and the company's name was changed again, this time to the Greyhound Corporation. At the same time, the Southern Pacific Motor Transportation Company and Pickwick Stages were joined in Pacific Securities, Inc., changing its name to Pacific Greyhound. That company then absorbed Yelloway-Pioneer Lines.[61] Thus did Greyhound evolve through a maze of owner relationships, all railroad oriented. For its part, Trailways was, and remains today, a confederation of independently owned and operated bus companies.

The editors of *Bus Transportation* noted, "The merged [Greyhound] operations represent a rolling stock equipment of 1,400 buses, covering 28,000 miles of route. These vehicles have a daily schedule of 190,000 miles. Ten million passengers are carried yearly, while it is estimated that a total of 73,000,000 bus miles are run up annually."[62] Ownership patterns set operating patterns, each ownership entity operating its own buses, establishing its own terminals, publishing its own timetables, and so on. It was not the best of times to launch a nationwide bus system. Severe economic depression, coupled with ever-increasing automobile use, had brought a serious decline in bus ridership. The enterprise was threatened from the beginning with bankruptcy, if not liquidation. Thus did the General Motors Corporation, manufacturer of Yellow trucks and buses, assume much of the Greyhound Corporation's debt, keeping the consolidated bus system open. Greyhound did fill an important transportation niche. It enabled Americans who could not afford cars—or, for that matter, expensive passenger-train accommodations—to travel. Before Greyhound, long-distance trip-taking by bus generally left much to be desired. It was not just the frequent changing of buses from one line to another but the general lack of quality vehicles and facilities. That changed.

In 1929 investigators were sent out by the Trans-Continental Passenger Association to ride buses across the country. Their findings were not encouraging. *Railway Age* reported, perhaps with a little bias:

The coaches are crowded and more attention seems to be paid to local than to through travel. . . . Apparently no attempt is made to regulate the heating of the coaches and frequent complaints are registered by passengers that the vehicles are either too hot or too cold. . . . Generally speaking, the roads are fair and in some places very good. In many instances, however, the coaches are operated over dirt roads, making the travel rough and uncomfortable. . . . Restaurants and eating places recommended by drivers are usually third-class and passengers are forced to shift for themselves. . . . Many restroom stops are made, but in the majority of cases the transportation company does not operate the restrooms and passengers are forced to use the facilities of hotels, garages, barber shops, etc. In some cases even railroad stations are used.

Although traveling by sleeper bus from Chicago or from St. Louis to California provided an interesting experience, all the investigators felt that continuous traveling night and day in such confined conditions put undue strain on travelers.[63] Greyhound, however, began to work hard to reduce such strain. A key improvement was the establishment of bus stations at logical break points along long-distance routes, providing travelers with comfortable lounges and quality restaurants. Brought to the fore were the Greyhound Post Houses and a subsidiary formed to build and manage them, although many came to be operated under individual franchise agreements. Coaches were upgraded, and the servicing of coaches was fully rationalized with refueling stations and garages carefully located throughout the consolidated system. Driver training was improved. A national advertising campaign was launched to reeducate the public regarding bus travel. Greyhound meant to provide a first-class travel experience but at a reasonable price.

Two railroads, the Great Northern and the Southern Pacific, played the key roles in establishing the Greyhound system. The Chicago, Burlington & Quincy played a minor role through a small stock holding in Pickwick-Greyhound. However, other railroads quickly moved to join the action, prime among them the Pennsylvania Railroad and Union Pacific's Chicago & North Western's Interstate Transit. The Pennsylvania in 1927 had turned over management of its lines to the Motor Transit Company. When Motor Transit became Greyhound, the Pennsylvania's bus properties became Pennsylvania-Greyhound.[64] Pennsylvania Greyhound held controlling interest in Illinois Greyhound and Indiana Greyhound. The jointly owned Union Pacific and Chicago & North Western Interstate Bus service had led Union Pacific to acquire several Pickwick-Greyhound routes, including the "Yelloway" routes. The combined operations were renamed Overland Greyhound (later merged into Central Greyhound).

The Boston & Maine Transportation Company (along with the bus operations of the Maine Central) was also brought into the Greyhound fold, forming Eastern Greyhound Lines.[65] The New Haven's bus subsidiary, the New England Transportation Company, was quickly added. Although it had not earlier involved itself directly in busing, the New York Central Railroad specifically formed Central Greyhound to enter the market, purchasing previously independent bus companies in seven states to do so. That operation spread from Boston to Chicago. The Richmond, Fredericksburg & Potomac Railroad held controlling interest in the newly formed Richmond Greyhound. Southern Kansas Greyhound combined the routes of the Santa Fe's Southern Kansas Stage Line with those of the bus subsidiary of the Frisco.[66] Santa Fe Trails, on the other hand, became part of Trailways.

National Trailways Bus System was organized in 1936, its principal organizers the Burlington Transportation Company, Santa Fe Trails, and the Missouri Pacific Transportation Company (fig. 8.7). The Interstate Commerce Commission strongly encouraged its organization, so dominant had Greyhound become in the long-distance busing field. The result was a bus association overseen by a central office charged with setting operating standards, coordinating schedules, and promoting nationwide advertising. It also played a role in designing bus equipment, thus making Trailways' long-distance operations distinctive. Over one hundred bus companies ultimately joined the association, each with "Trailways" inserted into its name. Yet, each remained independently owned and operated. As *Railway Age* reported, "While the buses owned by all companies will be painted the same color and bear the emblem of the National Trailways System, they will continue to be owned by the participating companies, Trailways merely handling national advertising and coordinating schedules, and the establishment of provisions for continuous passage on one ticket from any point on the network to another. Joint use of terminal facilities is being arranged. . . . New terminals are planned for New York, Chicago, Los Angeles, Cal., and other cities."[67] The system's operation was directed by a "managing committee" headquartered in Chicago.

Burlington Trailways was the initiator (fig. 8.8). The Burlington Transportation Company originated in 1929 when the Chicago, Burlington & Quincy Railroad purchased a local bus company operating between Peoria and Galesburg in

Illinois, the latter city an important junction point on the railroad. It spread its services first across Iowa and Nebraska. But by 1935 the company was operating in twelve states with through service from Chicago to the Pacific Coast.[68] Next in importance came Santa Fe Trailways organized around the Santa Fe's "Indian Detour" network centered in Santa Fe, New Mexico. The name "Trailways" was derived from that company's previous name: Santa Fe *Trails.* Then the former Missouri Pacific Transportation Company joined with routes extending south and southwest from St. Louis to New Orleans and into the Rio Grande Valley of Texas as well as west to Kansas City, Denver, and Salt Lake City. Various other railroad-controlled bus companies also joined; for example, one line previously

FIGURE 8.7. Advertisement for Burlington Trailways. *Burlington Trailways Time Table,* June 3, 1936, front cover.

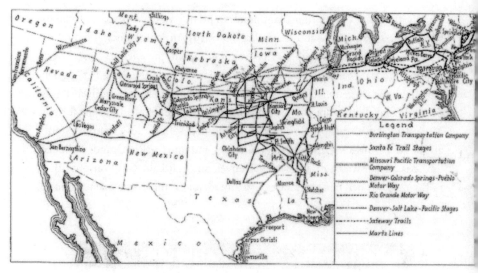

FIGURE 8.8. The National Trailways System, 1936. "National Trailways Begins Operations," *Railway Age*, n.s., 100 (Apr. 25, 1936): 693.

operated by a subsidiary of the Denver & Rio Grande Western became Rio Grande Trailways. Two other Rio Grande Railway routes, when linked with former Burlington operations, became respectively the Denver–Salt Lake–Pacific Trailways and the Denver–Colorado Springs–Pueblo Trailways.

BUS STATIONS

Buses, even so-called express buses on long-distance routes, stopped to drop off and pick up passengers when requested to do so along highways. In small towns, loading and unloading of passengers was usually accomplished at curbside, an adjacent storefront business providing ticketing and baggage-claim services as well as space for passengers to sit while waiting. It might be a hotel, a theater, or a drug store. As pictured in Rock Springs, Wyoming, in the 1930s, a main-street café served that purpose (fig. 8.10). The postcard's caption indicates that the café was "open all hours." In cities, however, curbside loading added to traffic congestion. An article in *Railway Age* warned in 1929, "Congestion is becoming a major issue in all of the larger cities, and to overcome this to some extent it will probably be necessary for the municipal authorities to eliminate curb stops, at least in restricted downtown areas, in so far as inter-urban motor coach operations are concerned, so that off-street terminals will have to be provided."[69]

The problem of curbside congestion was especially acute for the Baltimore & Ohio, Jersey Central, and Reading bus subsidiaries operating into and out of Manhattan. Thus was the Central Union Bus Terminal built, housed in a twenty-five-story building along with the Dixie Hotel on West Forty-Second Street near Grand Central Station. Not only did bus passengers have access to the hotel's amenities, but they had their own facilities, a waiting room, smoking room, lunchroom, confectionary store, "baggage and parcel" room (with storage lockers), a newsstand, an information bureau, and a bank of pay telephone booths. An interesting feature, one copied in many other big-city bus terminals, was the turntable for reorienting arriving buses for eventual departure. As *Railway Age* explained, "A Motor coach coming down the entrance ramp drives directly to a turntable, from which 10 loading stands radiate like tracks of a locomotive roundhouse. A coach leaving the terminal backs on to the turntable, is swung around so that it is heading for the exit ramp, and is driven out."[70] It was the Greyhound System, however, that led the way in new terminal design. As their buses (most of them built by General Motors) became increasingly streamlined

FIGURE 8.9. Greyhound bus terminal, Charleston, West Virginia, circa 1940. Postcard, authors' collection

FIGURE 8.10. Western Café, Rock Springs, Wyoming, circa 1935. Postcard, authors' collection.

(thus keeping up with the new streamlined passenger trains), new Greyhound stations across the country were styled in 1930s "Streamline Modern" also. Pictured is the new Greyhound terminal in Charleston, West Virginia (fig. 8.9). Lookalike bus terminals deliberately styled as being "modern" greatly added prestige to the Greyhound brand. It was a clear example of what has come to be called "place-product-packaging."

America's steam railroads participated fully in the first two phases of the bus industry's rise to maturation. At first, busing, like the initial operations of the electric interurbans, involved short-distance travel, mainly connecting big cities with outlying small cities and small towns. Again, busing brought to the railroads important cost savings as steam locomotives with standard passenger cars and even rail motor cars were replaced. But it was short-distance, line-replacement busing that truly brought the railroad bus subsidiaries into existence. Their buying-up of preexisting bus lines had much to do with eliminating competition but also with hitting the ground running—obtaining already viable bus routes without the delay of obtaining approval from state public service commissions. It was the private automobile, of course, that undermined railroad passenger ridership rather than buses, although competing bus lines, when they came to the fore, did not help. Growth of private automobile ownership and the popularity of motoring was the main culprit. And that undermined travel by bus as well as by rail, just as it had undermined ridership on interurbans and streetcars.

Declining short-distance bus patronage is what forced bus-line consolidation. As bus technology improved and as highways improved, long-distance busing became increasingly feasible. And it remained less costly to provide, although travel amenities were not quite as good as those of rail travel, a function mainly of bus passengers being more confined for longer periods of time. Here was opportunity, however, for the railroads to hedge their bets, if not cash in their chips by encouraging busing as a whole new industry. Short-distance bus lines could be amalgamated into long-distance systems, and the railroads could do so by buying out or consolidating companies and profit from doing so. By the mid-1930s, it was apparent that bus subsidiaries were more a hedge against the future than an essential aspect of railroad profitability. But then there were also the new streamlined passenger trains driven by more efficient and less costly-to-operate diesel locomotives. Perhaps railroad passenger service would revive? Perhaps busing would not be so important after all? Let Greyhound and Trailways alone assert future bus initiatives. And that, of course, is what happened. However, in 1939 American railroads were still variously involved with busing. Some fifty-two hundred railroad-controlled buses were still operating on more than fifty thousand miles of highways.[71]

Chapter 9

RAILROAD TRUCKING
SUBSIDIARIES

Trains and trucks—the two seemed so ill-disposed toward each other in the 1930s that it might have been hard to appreciate that they had actually benefited each other. With trucks in competition, railroads worked best for those with heavy freight to ship over long distances. Some railroad lines, however, wanted more. Good roads were no longer an advantage as they had been when they were built parallel to railroad rights-of-way. Networking roads gradually built for other services hurt railroad profits, causing railroads to complain about the truck's right-of-way, for which railroads wanted trucks to pay what they believed was their fair share. In this chapter, we trace the advent of the truck through railroad subsidization of their own truck lines. One must remember here that trucks were more nimble than railroads. They provided better less-than-carload and short-haul service than trains, which found them very competitive. Railroads had come into competition with the upstart truck for many types of transportation by the 1910s, but by the early 1930s trucks gave railroads a sense of unfair competition. The railroad trade journals grasped facts relevant to these circumstances but also carried the outcry that charged the narrative of rivalry.

TRUCKS

Still in the truck's introductory phase, the journal *Freight* worked in 1911 to per-suade people potentially interested in the truck that it could not be "passed by with indifference."[1] Three articles in that periodical early that year claimed that trucks were showing more and more that they were more economical to oper-ate than horses for a host of interlocking reasons. They enabled a larger scale of delivery after goods were off-loaded from trains and delivered to the consumer.[2] Linking railroads with trucks still further, it was advised to think of the truck as a derivative of the locomotive.[3] Their ruggedness was an asset because vehicles making deliveries had to navigate the commonly rough settings around both freight houses and on highways, and they were reliable. In a much longer article, the *World's Work* repeated these arguments and added that in cities the truck meant "cleaner streets and less congestion; for the suburbs, service that was be-yond the horse-drawn radius."[4]

During World War I, pressure on efficient delivery mounted. A highway transportation committee chaired by Roy D. Chapin of the Hudson Automo-bile Corporation, included an expert on good roads, a highway engineer, and an executive of a hauling company to determine whether trucks could be used for deliveries in congested terminals. The committee believed that trucks offered the wartime advantage of reducing the amount of coal without which railroads could not function.[5] Emergency thrust the truck into the forefront (fig. 9.1). Pierce-Arrow advertised that its motor trucks were adaptable, taking over for many horses: "In the widest variety of haulage problems, ranging from the traffic of peace to the transport of war, Pierce-Arrow fleets have proved their ability to meet the most difficult conditions of service," performing "quicker, better and cheaper" than horses and eliminating "the intolerable delays of freight conges-tion and embargoes by quickly and economically handling shipments formerly made by railroad."[6] They were, along with "motor buses," excellent as feeders to railroads, making branch railroads unnecessary and transferring short hauling to highways; and though thought to make railroads more efficient, they were not yet intended to put railroads out of operation. Nevertheless, the National Automobile Chamber of Commerce executive who declared those beliefs also included an apparent warning: "The big, national motor-truck era is just begin-ning."[7] While it was reckoned in 1911 that horses were not doomed to total re-placement,[8] this mode of transport came under scrutiny during the war because they were so comparatively costly.[9] Amid the enthusiastic roar in favor of the truck during the war, an executive of the Goodyear Tire and Rubber Company held that a truck with pneumatic tires would go faster than one with the currently popular solid tire and that this would enable it to compete with railroads. As if history was inherently on the side of trucks with pneumatic tires, the Goodyear executive preached: "The same forces which operated in favor of the railroads, as against the rivers and canals, are now operating in favor of the pneumatic-tired motor truck as against the railroads. They are principally speed, flexibility, and reduction and number of transfers, followed in many cases by a lowering of costs." Aside from the obvious advantage this implied for his products, the Goodyear executive struck a cautionary tone from transportation history by noting that the truck, too, would eventually be surpassed by the airplane because it was speedier.[10]

The Transportation Problem

THE Railroads *must* be relieved—
Goods *must* be moved—The wheels
of industry *must* be kept moving—

The Solution of the Problem lies in the use of
Motor Trucks for all hauls of one hundred miles
or less—Wilson Transportation Engineers will help
you in the adaptation of Wilson Trucks to *your*
transportation needs.

FIGURE 9.1. The Wilson truck manufacturing company put the urgency of truck use very graphically eight months into World War One. "Wilson," *Literary Digest,* 55 (Dec. 29, 1917): 96.

Motor trucks made great gains during the war. Railroads could not handle all the freight required; trucks were necessarily pressed into service.[11] Whereas passenger-car production dropped by more than 800,000 vehicles between 1917 and 1918, trucks jumped 76 percent, from 128,157 to 227,251.[12] The industry was poised at war's end to manufacture 75,000 more commercial trucks than before, since they would no longer be necessary for military purposes.[13] Moreover, railroad lines amounted to 350,000 miles, their potential inherently less than that of trucks because the latter had 2,205,000 highway miles on which they could serve.[14]

The truck figuratively bolted out of the war with a most promising future. One contemporary analyst would look back upon 1917–1920 as a "boom period."[15] Not only all but replacing horse transportation, it also took over from the railroad in certain circumstances. Some railroads substituted trucks to serve on spurs and short hauls (one hundred to three hundred miles), thereby enabling railroads to concentrate on essential freight over long hauls. Trucks also proved to others that it overcame the problems of slow local freight trains but could also serve better for some long hauls. In most long-distance shipments, however, trains

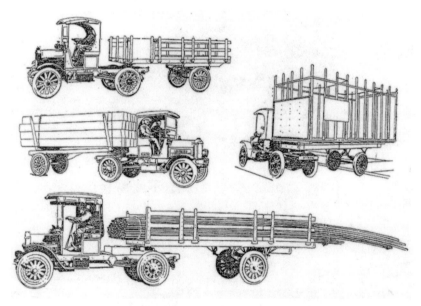

FIGURE 9.2. Early variations of the semi-trailer, known in the early-1920s as the "semi-tractor," could haul cargo loaded in several ways. Victor W. Page, *The Modern Motor Truck: Its Design, Construction, Operation and Repair* (New York: Norman W. Henley Publishing, 1921), 589.

still functioned better because labor costs were lower. Freight trains could move three thousand tons with six crewmen, whereas it would require six hundred five-ton trucks and at least six hundred drivers and maybe as many helpers.[16] Six years after trailer production began, just before World War I, advocates spoke for its potential to increase trucks' freighting capacity (fig. 9.2).[17] (This article originally appeared in *Commercial Car Journal.*) A switchman's strike, *Highway Transportation* asserted in 1921, very nearly caused rail shipping to end, thereby revealing trucks to be favorable by contrast.[18]

Railroads pushed for coordination with trucks, not the triumph of the latter. "Terminal trucking" emerged in trade jargon of the time for the harmonious meeting of trucks and trains. Trucks picked up and delivered less-than-carload quantities at railroad stations. The scale of the fortunes at risk being so great, another railroad spokesman assumed a prophetic tone: "Motor vehicles and concrete highways are not destined to erase the railroads from the map, but are destined to add greatly to the utility and efficiency of the railroads by rounding out and completing the rail service, and relieving the railroads of certain forms of service which are burdensome."[19]

Government would be required for the truck's fullest potential to be realized. Highway construction and maintenance were essential.[20] Amid the truck's rapidly rising popularity in the early 1920s, the competitive *Electric Railway Journal* pronounced that regulation of some truckers' erratic behavior, such as refusing to solicit freight for a full carload on return trips, would also be necessary.[21] In 1923 a committee devising a comprehensive plan that would tie truck and railroad transportation together in maximum productive harmony reported on ways to ease cities' congested terminal areas; these would include relieving railroads of uneconomical services via truck service, plus more use of self-propelled railway cars and supplementary bus service.[22] Railroad managers were concerned that the place of motor vehicles as common carriers, including trucks, was being left undetermined by federal and state commissions. The truck was neither taxed nor regulated "commensurate with the privileges it enjoys in the use of the highways," a railroad executive wrote in 1924.[23] People with public responsibility for transportation encouraged those in the transportation business to start thinking about how and to what ends they should direct motorized transportation for the most desirable effects on life and business.[24]

Trucking managers fought back against charges that their companies took advantage of circumstances to achieve growth. Railroads complained about paying for highways from which their motorized competition profited even as statistics proved they paid less than motor vehicle owners. Also, while trucks paid 10 percent of their gross revenue for public highway use, between 1911 to 1922, railroads charged an average of 13.3 percent of their operating income for the maintenance of way and structures.[25] An executive of the White Motor Truck Company charged that "for some years attempts have been made to smother the commercial motor vehicle with regulations and to throttle it by taxation."[26] Squabbling aside, the railroad-prone *Railway Age* used a questionnaire to show that twenty-four Class I railroads operated over eighteen hundred trucks, tractors, and trailers in 1927.[27]

Trucks' commercial potency was not only reflected in this give and take but in raw numbers. In 1928 the tenth edition of the National Automobile Chamber of Commerce's automobile industry statistics showed that truck registration rose from 410 in 1906, the first year of registration, to 2,896,886, and that there were 272,365 trucks in motor truck fleets of two or more, 105,717 in motor truck fleets of three or more, 37,277 in motor truck fleets of five or more, and 10,663 trucks in motor fleets of ten or more.[28]

Trucking continued into the 1930s along lines already well laid out in transportation trade and popular magazines. What should trucks pay for their sine qua non, highways? Should states levy special taxes beyond registration fees and a gasoline tax? Twenty-three states did so. By 1929 two states levied a flat rate, ten states levied an "actual use" tax, fourteen states levied a potential use tax, and four states combined an "actual" with a potential use tax. Trucks paid about 5 percent of their annual income while railroads paid about 12 percent of the operating revenue only to maintain the right of way.[29] *Railway Age* went so far in 1933 as to call for an end to federal highway aid appropriations,[30] also hastening to point out that two-thirds of the states' indebtedness resulted from their highway policies.[31] The trade journal stated bitterly that a "single cause of such a vast increase of government indebtedness and expenditures raises economic questions too important to be ignored when as is actually the case, propaganda is being industriously carried on to keep the cause in operation."[32] The form of transportation seemed to be outside most laws and regulation at the same time that its ton miles increased and those of trains decreased.[33] Did trucks take away railroads' earning power? This stemmed from trucks increasing their long-distance service, and the question might not have been asked if the Great Depression were not underway.[34] New to data-collecting among truckers was the complaint that data gathering was difficult, perhaps another worry due to the challenged railroad industry in the 1930s.[35] Arguments again came forth from the public for finding a place for both railroads and trucking. What the *Saturday Evening Post* wanted was an answer to "whether they are to pull together or in contrary directions."[36] Railroads were struggling; trucking was almost perceived as their enemy. Trucking accused railroads of trying to put trucking out of business.[37] Yet, trucking interests gladly pointed out that they could help railroads at terminals. These facilities were congested, but trucks helped diminish that by their shorter handling times. They also lowered rates for railroad labor and equipment.[38]

A cooperative spirit eventually entered the scene. In 1940 a report by the Automobile Manufacturers Association stated that between 1921 and 1937 motor vehicles had paid well in excess ($385,000,000) of their usual contribution for highways and roads.[39] Trucking technology in the 1930s also had worked further toward lower costs with the introduction of the diesel engine for motive power.[40] At the end of World War II, the Association of American Railroads outlined those services that railroads could best offer the public if they used trucks and buses: pickup and delivery, intraterminal transfer, substitution for line-haul rail service to short-haul areas around larger gathering points, and substitution for other railroad-line haul.[41]

LOCAL DELIVERY

A full decade into the twentieth century, motor truck advocates still needed to persuade many about their mode's superiority over horses, depending on the distance of delivery. Over twenty miles, trucks were certainly better than horses, but they did demand attention regarding the driver's capacity and supervision, and the truck as a machine. Horses were still competitive in 1911 within shorter distances. Summer and winter deliveries affected the relative differences too. Trucks could be overloaded beyond their capacity with negative consequences, while horses always seemed to have "reserve capacity." A two-horse vehicle was comparable at the time to a three-ton truck. In this narrative, the trade journal *Freight* admitted in an editorial that the motor truck in its "early stages" had been a weak link in the transportation chain. "However," it added, "motor trucks have been greatly improved since their introduction, so much so that some manufacturers of these vehicles are now advertising their products as representing the last word in perfection."[42] Ninety percent of all railroad freight was still being handled by horses at one or both ends of the shipment. Those determining whether to use horses or trucks also required education about the two.[43] Those considering one or the other transportation means should know that it was cheaper to maintain trucks than horses and that they saved time, advised one transportation professional. The Bush Terminal Company operating in South Brooklyn established a delivery system founded on the relative advantage of trucks but also required construction of a ten-story freight station to which freight was delivered for transfer to a series of outlying, smaller stations in New York City.[44]

Motor truck shipments were far less expensive than railroad freight. A key difference was the higher cost for packaging freight for railroads, as contrasted with trucks, because railroad shipments required more handling. Trucking required protection only for loading and unloading.[45] Articles unfavorably contrasting the cost of horse-versus-truck transportation continued into the early years of the twentieth century's second decade.[46] Statistics gathered in Massachusetts revealed the growth of local motor-freighting in 1912 and 1915: on the road from Boston to Quincy, 35 versus 146 trucks; Fall River to New Bedford, 29 versus 230 trucks; and Worcester to Boston (via Shrewsbury), 15 versus 45 trucks.[47] Pressures from World War I induced the Merchants' Association of New York to develop a "store-door" delivery system requiring trucks to make the final stop after taking freight from railroads.[48] Upon the government's encouragement that automobile lines assume short hauls from existing railroad services, the treasurer of a trucking company organized the O'Connell-Manly Truck Company in Waukegan, Illinois, serving a distance of thirty-six miles.[49] A transportation engineer for the Packard Motor Car company outlined extensive, statistically supported explanations for the superiority of trucks over railroads in short hauls. Much of the explanation echoed reasons previously stated, but the engineer's significant conclusion was that statistics were increasingly important in the trade and "as the railroad short haul runs up into the millions of tons it will be seen that the possibilities for the use of the motor truck in this field are tremendous." He prophesied that as statistical analysis became more common as a basis of decision making, "eventually all transportation [would] be sold on that basis, and because of this alone the motor truck [was] going to find a constantly-increasing market."[50]

Agricultural commodities benefited too; cotton, for example, was efficiently delivered from the fields to local compressors, either by train or truck, but trucks were used exclusively to freight the bales to the railroads. Trucks were also driving the four-mule "cotton float" out of business in moving cotton on the Mississippi

River levees (fig. 9.3).[51] A milling company forty miles west of Chicago operated its own truck profitably despite nearby interurban and steam railroads.[52] A garage superintendent, a year after the above-mentioned argument that statistical proofs would lure more truckers into the short-haul market, fully agreed.[53] A United States senator from Kansas gave a wide-ranging speech in 1920 in favor of trucks for farming and encouraged manufacturers to place more motorized vehicles on sale in the Middle West, where farmers were "the quickest to buy any necessity or luxury that the market affords."[54] Trucks also advanced in livestock shipment to stockyards through the decade so that, by 1929, 23 percent of the total receipts at seventeen markets in Chicago, Omaha, Nebraska, Kansas City, Missouri, and St. Louis came from truck delivery.[55]

Economic events in the early 1920s highlighted the potential benefits of local trucking. A severe winter and a railroad strike in 1919 added impetus to trucking, one observer in Connecticut, Pennsylvania, and New York reported.[56] As a cure for the doldrums, *Highway Transportation* lectured at the end of 1920, "We have reached the school grade that teaches, 'Motor trucks for short hauls!'" But it also called for more terminals and markets to handle the produced abundance.[57]

Local delivery gained further momentum as the 1920s unfolded. So convinced was the Chicago, North Shore & Milwaukee Railroad on principle of the contribution that a station would make in keeping trucks working while freight

FIGURE 9.3. Short-haul trucks at the Riverside Mills, Atlanta, Georgia. "How The Motor Truck Fits into the Cotton Industry—Plantation And Mill," *Freight Transportation Digest*, 1 (Nov. 1919): 5.

was loaded onto or taken from trailers that it built and operated from a station without knowing how much business would actually be received without the facility. *Electric Railway Journal* published an article in 1921 about this exemplary operation.[58] A year before, *Freight Transportation Digest* had run an article about the Detroit Creamery Company's daily collection of raw milk brought by truck and train from within a seventy-mile radius for processing at the creamery and the finished product's delivery via 107 motor trucks throughout Detroit.[59] *Railway Age,* only a few months later, argued for local delivery—"short haul" in the trade jargon of the day—if it was properly coordinated with the railroad; it concluded that a federally controlled, nationwide highway system should be put in place to increase truck implementation.[60] *Railway Age* editorialized in favor of trucks handling freight at outlying stations and assembly points to relieve the overloading at railroad terminals.[61] *Freight Transportation Digest* reprinted a *New York Times* editorial claiming that the railroad and the 1.5-ton truck could compete in a range of 180 miles and a 6-ton truck with the railroad in a range of 360 miles.[62]

Railroads increasingly took on means akin to trucks and developed their own short-haul trade. In a *Railway Age* article, a railroad executive claimed that, in handling packaged freight from terminals, tractors with trailers could provide more rapid and less costly service.[63] To stay in competition with trucks, more railroads were encouraged to reduce their rates.[64] Just as trolleys took away the railroads' passenger service over local distances, railroads were encouraged to learn from this that they should use motor trucks to transport raw materials to nearby manufacturing centers. Railroads were "national highways," the author said in trying to clinch his argument.[65] More and more fashioned for automobility, railroads were forsaking their historical functions. A Southern Pacific executive proposed that "highways serve as feeders for the railroads and develop traffic for them." An editor of the *Washington Herald* put it stoically: "The railroads will have to readjust their operations largely to short-haul and to high-class traffic. On this basis both the waterways and the motor-trucks will be traffic-feeders. They will also materially develop the territory the railroads serve and so increase high-class freights which can stand high rates."[66] *Railway Age* mentioned an unidentified western railroad that developed its own system of concrete roads used for transferring materials in upgrading its central storage yard.[67] In 1932 the Wabash Railroad successfully experimented with the "any quantity" plan at several west-central Illinois terminals, whereby truckers could unload any quantity of livestock for the railroad's standard rate, but only on Tuesdays. Thus, the truck ended the carload as the standard unit of measure in this successful experiment.[68] Other railroads tried their own variants of local truck shipping.[69]

However, the editor of *Railway Age* admitted in 1926 that serious cost analysis of motor-truck (as well as motor-bus) usage had yet to be undertaken and promised a series of articles "to open up the matter."[70] Through the late 1920s to the dawn of World War II, the trade literature recorded the railroads' increasing use of trucks for short hauls. An electric railroad in southern California used its own and the area's many privately owned trucks for short hauls, and *Railway Age* showcased the innovative work. It asked, too, if statistical analysis would determine the basis of decision making: "Possibly the ultimate success of this new type of co-ordinated rail-truck operation may be more reliably indicated by this popular demand for its expansion, rather than by the immediate profits being made from it."[71] Data gathered over a decade later reported that truckers for hire profited but not to a great degree, many making less than normal wages.[72]

Only rarely did the trade literature mention the itinerant trucker, the "gypsy trucker" in the jargon before World War II. Was the gypsy trucker a symptom of change in the trucking market? *Railway Age* asked this question in 1939 before describing the itinerant's work. They eluded all taxes except what they paid for gasoline and their truck's registration. They operated largely in agricultural areas where they went directly to farms and bypassed elevators. They operated in the local delivery market. *Railway Age* surveyed the market because it received so many complaints about the gypsies; an estimated six hundred elevators went out of business in eight western states because of these truckers. Lumber and coal merchants were also hurt by the gypsies. *Railway Age* estimated that these freelancers would grow in number just as surely as they operated with the great chance of going out of business.[73] This segment of trucking was incredibly flexible.

Economy and flexibility—the railroad could never offer these transportation virtues to the degree that the truck could. Ironically, only by using the truck for local delivery and less-than-carload freight after arrival at a terminal could the railroad function more economically and flexibly.

LESS-THAN-CARLOAD FREIGHT

To make them popular, trucks were pressed into use as interchange services for railroads from the time they became reliable for commerce. Taking the form of transfer between railroads, local package freight was reported in 1920 to exceed more than 4 percent of the total that railroads shipped. This freight could pass between one freight house to another or directly across a city, either by trucks or "trap cars," the colloquial term for railroad cars that transferred freight across a town between railroads. Railroad terminals suffered from insufficient capacity by the end of World War I. Anything that increased the amount of freight that was passed through terminals improved railroads' capacity, and trucks enabled this improvement. "Trap cars" did not improve conditions. Trucks saved time, too; in Cincinnati they saved fifty-two hours in the average interchange. These results also meant cost savings to those sending and those receiving. "In view of these showings, the motor truck warrants careful consideration from operating officers," *Railway Age* opined.[74] In Bridgeport, Connecticut, "one of the busiest manufacturing cities of the country," an experiment was conducted in 1920 to see how much speedier service could be if local manufacturers notified railroads a day in advance of loading that they had less-than-carload (or LCL) freight; it also enabled factories to continue production after the quantities for given cities were moved to the railroads, thus not crowding production space required for storage. The experiment satisfied railroads and manufacturers alike (fig. 9.4).[75] At the Chicago Junction freight house, a system was worked out whereby freight at the receiving door was put onto four-wheel hand trucks, thereby keeping the freight on wheels from arrival until storage. Each four-wheel truck was preassigned for a particular railway freight car handled by a particular group of workers. Although hiring labor to handle the freight was expensive, this was outweighed by the four-wheel trucks' carrying capacity. LCL freight was rendered even better, more efficient, and profitable.[76] Later in 1920 than the improvements reported above, Cincinnati implemented a speedier and cheaper system in the use of freight prepackaged in special containers.[77] New York Central introduced the container car, a rail car with nine separate containers.[78] One analyst taking a long-term historical view held that trucks were best in distances of about fifty miles for the less-than-carload sector of trucking.[79] Problems would eventually arise with various new schemes put into practice, but *Railway Age* tried to assure railroads that handling LCL freight would yield great benefits.[80] By 1924 it became the talk of the day in transportation circles.[81] In 1928 the Boston & Maine's highway subsidiary began a variant that utilized the subsidiary's trucks to store in local terminals LCL freight that other companies had previously handled.[82] Trucks were seen as revolutionary and no longer competitors by many railroad executives. The vice president of the Boston and Maine Railroad conceded, "Within their own field the trucks are unbeatable."[83]

FIGURE 9.4. Less-than-carload freight required pickup and delivery service ideal for trucks to handle. Denver & Rio Grande Western Railroad timetable, June 11, 1939, 4.

LONG-DISTANCE TRUCKING

Exploration of the possibility of long-distance trucking did not dawn until the second decade of the twentieth century. A rare case was the five-ton "Pioneer Freighter" of the Saurer Motor Company of New York, which in 1911 ran three and a half tons from Denver to San Francisco and back to New York City. Navigating over a collection of many poorly developed roads, the record truly sounds like a pioneering feat. Exemplifying long-distance trucking's novelty at the time, an "automobile truck," as trucks then were called, made news traveling in one day over 220 miles between New York and Philadelphia.[84] In California, the cradle of so many of automobilty's innovative uses, an interurban railroad attempted to bolster its sagging railroad business by utilizing a motorized truck that could run off and on rails.[85] Trucking's unquestioned capacities periodically made news but often with the caveat that it should pay its share of highway construction and maintenance. As one reader of the *Railway Age Gazette* commented to the editor about trucks paying their rightful share, it had to be remembered that "motor trucks [could] help railways by short hauls to relieve cars for long hauls, especially in intra-city switching movements."[86]

With America's entry into World War I, long-distance trucking made great strides. Its significance extended "from the interior of the United States right to the battle fronts of France and Belgium," in the words of *Highway Transportation*.[87] In December 1917 thirty army trucks ran in a drivers' training mission from Detroit to the Atlantic seaboard, traveling over six hundred miles. Some thirty thousand trucks were to eventually follow and, in the process, help train drivers for France.[88] The War Department further tested the potential in running thirty-three trucks from Detroit to Newport News, Virginia.[89] Calling these trials successful, the United States Quartermaster's Department decided to run "trains"—that is, convoys—every day for six weeks in early 1918.[90] Trucks with trailers came into commercial use.[91] The Post Office Department started to extend truck-driven parcel post routes in 1918.[92] "Will Displace Railroads for Distances Up to 100 Miles on Good Roads," *Highway Transportation* titled an overview article in 1918. That year would "probably be pointed out as the year in which the motor truck came into its own," the trade journal insisted.[93] Truck production increased over the first few months of the year, and it was estimated that it would almost double by year's end from 1917's roughly 180,000, with a truck in Maryland running routes up to forty-seven miles.[94] In other cases, eighty miles was a long run.[95] Goodyear Tire and Rubber Company advertised pneumatic tires on which a truck could run from coast to coast in thirteen and a half days (fig. 9.5).

Trucking and railroad interests initially took two different courses in the early 1920s to advance their respective causes. Trucking interests believed their cause could be advanced further by building highways. *Highway Transportation* editorialized that a company in the San Joaquin Valley, California, "where the roads [were] well built" and business was sufficient to run on a regular schedule, showed how effective trucks could be in general "when the national system of hard-surfaced highways, now planned, [was] completed."[96] A professor of highway engineering and transport outlined the highway features needed.[97] *Railway*

Age, only three months later, claimed that the truck's capability of competing with the railroad's capacity in long hauls resulted only from the temporarily high prices of certain goods from which trucks could exploit high rates.[98] Countering one transportation engineer's call for "special motor truck highways," *Railway Age* went so far as to claim that a horse's service area in a day was more economical than a truck's.[99] Railroads were not necessarily hanging desperately onto an antiquated view. Pointing out a prestigious trucking company's loss of $14,000 over one year of long-haul trucking, the secretary of agriculture in 1925 generalized that trucks could not compete with railroads in long hauling: "There was a time, no doubt, just after the war when enthusiasts thought they could see the truck taking the place of the railroad completely—at least they talked that way. But that time is past; and the reason for its passing is that the long haul doesn't pay—and truck operators know it doesn't."[100]

A trucking company executive clarified the circumstances, pointing out that trucks in the earliest period could compete with railroads inside a hundred-mile range. When roads and automotive equipment improved, the truck's competitive mileage doubled and by 1930 it could serve well in a five-hundred-mile range.[101] An industrial traffic manager, who wished to remain anonymous, contended that in California trucks competed very well against trains because truckers would go to any extreme in lowering their rates to get business, even acting as "cut-throat competition of the most ruthless sort between the truckers themselves."[102]

FIGURE 9.5. Packard and Goodyear's record-setting coast-to-coast truck in 1920. "Cross-Country Record Breaker Designed For Pneumatic Tires," *Freight Transportation Digest,* 2 (Aug. 1920): 2.

As noted in above, the lines between railroads and trucks were no longer drawn so bleakly as they were in the early 1920s. Improved organization and management increased trucks' capability to compete with railroads for long-haul service, a *Railway Age* editor conceded in 1931, but he also observed the considerable degree to which railroads developed their own trucking services.[103] Discussions about long-haul services lapsed in discussions about railroad and trucking in the 1930s.[104]

Federal authority in 1938 took steps relevant to long-distance driving when the Interstate Commerce Commission imposed the rule that every trucker have eight hours of rest after driving ten hours. This was warned to be only an "initial step," as factual evidence already being collected at the time of the commission's decision would determine whether or not further restrictions were to be imposed.[105] Driver fatigue was at issue and long-haul trucking was obviously the cause, although not mentioned. Trucking long distances to "terminal cities" was not a consideration that small carriers could especially afford to consider, and small carriers were understood to often perform services unavailable from other transportation companies.[106] The railroads' abuse of labor had been notorious. Long-distance trucking took its own tolls on labor, something of which the trade literature made no mention.

RAILROAD TRUCK SUBSIDIARIES

Transportation's embryonic stage relative to railroads and various automotive types in the early twentieth century explains the exchanges that later settled into familiar patterns. Trucks had the potential very soon to crimp railroads' profit making and even viability altogether. In 1912 *Railway Age Gazette* recorded a pioneering example of a manufacturing company that gave birth to its own trucking company on Long Island to deliver its manufactured goods in order to avoid delayed railroad service. It made possible a shipment from a factory in the morning, its arrival at the distribution point in the afternoon, and spared the consumer a full day when it arrived.[107] All was not at cross-purposes, however. We showed above how railroads began cooperating with trucks for short hauling and also LCL. Here we make special note of trucking lines that railroads developed as subsidiaries—another of the many currents in early railroading and automobility.

One of the first railroad-subsidized trucking operations was that of the New York Railways Company begun in 1913 to substitute for horse-drawn wagons.

The company was an electric railroad that purchased twenty-nine of the period's largest electric trucks from a truck manufacturer.[108] The steam-driven Union Pacific Railroad was still among the pioneers when it purchased trucks for delivery in Omaha, Nebraska.[109] In 1917 the Cleveland, Cincinnati, Chicago & St. Louis railroad began shunting LCL freight on preloaded bodies by truck to substations. It was anticipated that the service would increase the freight house's capacity because movement would be continuous and hold down loss and damage liabilities.[110] In 1917 wartime urgency pressed the new Highways Transport Committee of the Council of National Defense to encourage manufacturers to hire trucks since railroads could not respond quickly, often bottling up freight in their terminals. The committee sponsored a newspaper advertisement in New York stating that wartime urgency could not countenance quibbling about slowed railroad service: "Let everybody be prepared in some way, somehow, to move their merchandise away from terminals immediately."[111] Literature for the general audience noted how war momentarily had spurred railroad and truck cooperation but also foresaw its continuation later as a matter of business efficiency.[112]

So it was that the two transportation types increased cooperation in the 1920s. Railroad yards began to be fitted with special equipment for handling trucks. A writer for a popular magazine wondered if the railroads' earlier discontent had not simply replayed transportation's distaste for competition.[113] The Erie Railroad's vice president chronicled his company's arrangement with one of the largest trucking companies of the time, the United States Trucking Corporation, to have it deliver freight on the New Jersey side of the Hudson River, thereby saving time for the customer and terminal expenses for the railroad.[114] This was an essential part—but only one part—of a plan to render the port of New York more effective and less expensive from which to operate, in contrast with other ports on the Atlantic coast.[115] The Pennsylvania Railroad operated in St. Louis with tractor trailers used to offload containers with freight and put them onto other railroads.[116] In 1923 that railroad's vice president, speaking in congratulatory terms to a meeting of the Society of Automotive Engineers for their members' work, recommended train-truck cooperation but advised that neither own the other. Not only would it further complicate railroads' already complicated work, but it also would likely beckon restrictive legislation, which separate ownership would not. He concluded, "There is only one further thought which I wish to leave you with. The spirit in which all of our problems of coordination ought to be approached should be one of live and let live . . . to guide the development of that relationship along sane and sensible lines, so that each form of transportation may be enabled to give the maximum service of which it is capable."[117]

A survey in 1924 indicated the considerable degree to which railroads acted as agents sharing business with motor trucks. In the United States and Canada, fifty-one railroads supplemented their shipping by using trucks. Fifteen railroads contemplated using more trucks or using them for the first time. Twenty steam railroads or their subsidiaries owned over 219 buses. Eighteen steam railroads that did not use buses considered their installation.[118] Expansion was predicted for railroad-subsidized trucks and buses.[119] By 1927 a national meeting of railroad superintendents received a report of ten reasons why railroads preferred to op-

FIGURE 9.6. KATY Lines of the Missouri, Kansas, and Texas Railroad extended freight service between merchants and the railroad. Missouri, Kansas, and Texas Railroad timetable, June 1, 1936, 1. Authors' collection.

erate with trucks via subsidiary companies more than with railroads or a motor truck operator. They added up to economic operation and faster movement.[120]

In the advancing use of trucks for business that railroads started, it also seemed that it would no longer be possible to keep trucks free of state and federally regulated rates and services and special taxes for road use.[121] Trucking, too, continued having to contend that it paid its fair share of the highways that enabled their business. *Railway Transportation* charged the trucking industry in many cases of not caring to place value on timely delivery, as contrasted with railroads, which always did so.[122] Trucking faced its own future problems as a result. Thus railroads struggling to maintain solvency adapted trucks for their use.[123] During 1929 trucking grew partly because railroads ordered 1,764 more motor trucks and 390 motor buses (fig. 9.6).[124]

Two economists' scholarly papers from the early 1930s reflect significantly on the issue of railroad truck subsidiaries. Implicitly they are documents of a level removed from the journalistic medium of the trade journals, which seldom viewed the subject from a long-term perspective. These papers also show how serious in another way railroad truck subsidiaries became in the 1930s. These guidelines to the discussion through that decade are a published economic thesis by Ping Nan Wang, *Rail and Motor Carriers: Competition and Regulation* (1932), and William Cunningham's "Correlation of Rail and Highway Transportation," an article published in the *American Economic Review* (1932).[125]

The title of chapter 5 in Wang's thesis, "Railroad-Motor Coordination," connotes that trucks and buses, on the one hand, and railroads, on the other, had suspended fractious exchanges in pursuit of harmonious relations in order to benefit businesses and the public alike. They had sorted through three options: (1) contractual services, which independently owned trucking companies preferred, (2) railroad-operated services, which were few and not extensive, and railroad subsidiaries, which railroads preferred.[126] The Transportation Act of 1920 had forestalled antitrust prohibition of these subsidiaries so long as they were judged to be within the public's overall interest.[127] Cunningham called for regulatory bodies to decide in principle what part of highway costs should be charged to trucking and what part to the public. Moving forward, Cunningham wanted the states to regulate licenses of operation and rates that truckers charged; because of the differences, "diversity and confusion" reigned (fig. 9.7).[128]

Federal authorities and transportation executives began to step into the contest in the name of general coordination between trucks and trains. President

From the Columbus (Ohio) Dispatch

FIGURE 9.7. An editorial from the *Columbus (Ohio) Dispatch* illustrating the common railroad contention that trucks competed unfairly with railroads because they did not pay their share for there right of way. *Railway Age*, n.s., 92 (Apr. 30, 1932): 724.

Roosevelt spoke in favor of unifying the regulation of all transportation.[129] The federal coordinator of transportation did so as well.[130] The Great Depression further irritated circumstances, with the railroads raising their new claim of unfair competitive circumstances because the railroads' labor cost more than truckers were willing to receive and resulted in an overall deflationary affect on the depressed economy.[131] Good railroad income meant a good economy that was important for recovery from the depression.[132] Before the Louisville Transportation Club in 1935, a speaker called on the trucking industry's "subordination of its own private gain, if this be necessary to promote the public good."[133]

Railroad and motorized transportation had achieved considerable coordination on the eve of World War II. In 1939 the Kansas City Southern Railroad stood out for having developed a trucking service that paralleled its entire track system. Also notable was the Illinois Central, which coordinated rail with highway service throughout the central and southern parts of the state. The Missouri-Kansas-Texas Railroad coordinated truck service throughout Kansas.[134] The Association of American Railroads assigned three committees to advise how service might be improved.[135] By mid-1940 the Interstate Commerce Commission entertained railroad companies, the railroad union, and some chambers of commerce in favor of ending restrictions that prevented coordination. Some railroad executives, however, still seemed restive about cooperation, as they testified before the commission, citing highly qualified circumstances under which it could be practiced.[136]

Thus, appreciating the operational strengths that trucks both displayed in maneuvering and yielded on the business ledger, railroads developed subsidiary relationships with trucks fairly soon after they came into operation. However, they did not universally transform the motor and railroad interplay all at once. Railroad truck subsidiaries grew from several regional locations. We turn next to describing the hearths from which this feature sprang before its acceptance nationwide.

RAILROAD TRUCK SUBSIDIARIES BY REGION

Northeast

Regions where railroads did not depend heavily on short-haul transportation entered the subsidiary field later. Short haul and subsidiaries, on the other hand, worked well together. The New England–based Boston & Maine was one of the first railroads to develop a truck subsidiary.[137] Operating in a comparatively small region with large manufacturing centers, the general manager of the New York, New Haven & Hartford Railroad stated that distribution via highways after railroad handling could be more efficient, equipment and facilities better utilized, and costs to individuals and communities reduced.[138] The Erie Railroad developed a system with consigned trucks and tractors, trailers, and ferries to lessen the congestion at the port of New York, likely in 1921.[139] In 1925 the Boston & Maine undertook a huge reorganization dependent on its Boston & Maine Transportation Company subsidiary, reducing the railroad mileage from 2,450 to 1,450.[140] This was not only because of the closeness of cities, but also because of the LCL traffic, the good highways, and the absence of state regulations.[141] In a sign of growing rationalism in this subfield of railroading and trucking, railroads turned over to the *Railway Age* a large volume of ordinarily very private financial data for a discussion of the subject. In 1926 the trade journal reported the Lehigh Valley Railroad's use of its own trucks to transfer loads between ships and trains, the Long Island Railroad's use of its own trucks and trailers, and in 1927 the Baltimore & Ohio Railroad's contracts with five truck companies for transfers on Long Island.[142] Other railroads followed suit through the 1930s, either increasing their work through a subsidiary or starting work with a subsidiary.[143]

Southeast

The Southeast's railroads first grew along the Atlantic coast in the mid-nineteenth century,[144] but when they expanded rapidly—twice as fast as the rest of the nation between 1880 and 1925—the expansion occurred along the Mississippi River valley.[145] Individual accomplishments within the region, however, waited until the 1920s and 1930s. The Norfolk & Western's terminal was enlarged from six thousand to thirty-seven thousand square feet in 1925, and the railroad's own trailers handled the commodities. Reliable data proved that, in what was no small feat, the cost of handling freight at the new station had dropped to forty-five cents per ton from the approximately eighty-three cents per ton at the old station.[146] Store-door delivery and LCL hauling boomed rather late in the Southeast, when in March 1933 eight railroads took to the systems and the railroads contracted for motor trucks to do the work. *Railway Age* carried a very detailed report of the new program, in keeping with the growing case for rational appeals.[147] Earlier developments on these lines occurred to save costs, but their latest spur was the desire for speed and flexibility. The year 1933 witnessed a rise in the latter two elements, possible proof of the increasingly serious quality of railroad management, as opposed to impulsive reaction always in favor of private profit and less for public satisfaction.[148]

Midwest

St. Louis witnessed the prelude to midwestern railroad subsidiaries in 1902. The Columbia Transfer Company began that year, opened a depot to receive freight two years later, and in 1915 purchased the St. Louis Transfer Company. The Terminal Railroad Association of St. Louis, which fifteen railroads entering St. Louis owned, maintained its own station and another at East St. Louis, Illinois, for freight receipt. In a step unreported in other trade journals, *Railway Age* noted that the move's effectiveness was due to the enlistment of employee cooperation, special education—most of all regarding freight's careful handling—and a bonus to stimulate productivity.[149] The Wabash Railway Company contracted with the Arthur Dixon Transfer Company in Chicago to haul freight, but the railroad loaded and unloaded the freight. Only because the railroad was inexperienced in trucking did it arrange for the contract.[150] In 1927 the Chicago, Rock Island & Pacific Railroad moved from a six-month experiment with its trucks to a subsidiary trucking line.[151] The Keeshin Motor Express Company, which was incorporated to work with motor carriers and had a contract with the Chicago, Rock Island & Pacific to move (or "ferry" in transportation trade jargon) its highway trailers across Illinois,[152] emerged by 1936 as a key holding company for motor carriers and car-loading companies and eventually worked with four prominent midwestern railroads.[153] Wilson Transportation Company in Sioux Falls, South Dakota, which was incorporated to haul freight throughout the state and nearby Iowa, had ten of its routes purchased by the Chicago, St. Paul, Minneapolis & Omaha, itself owned by the Chicago & North Western. The truck's ability "to save distance and time over well-established rail routes was the secret of the Wilson Transportation Company," its vice president assured an audience of the Motor Transport Division of the American Railway Association in 1930.[154] The Illinois Central Railroad took the strategy of contracting with the Railway Express Agency to coordinate shipping to sixty-six stations around Carbondale, a key southern Illinois distribution point. The recent discovery of the area's oil fields created a business opportunity. *Railway Age* ardently reported, "Not only has the flow of merchandise to local stations on these branches been naturally increased, but, as is the case in all oil fields, the receivers of the freight are invariably in a tremendous hurry to have it delivered."[155] Of all the locales, the Midwest seems to have experienced the most variations in whether truckers or railroaders initiated the subsidiaries to their mutual benefit.

West

Subsidiary service within the West—but not tying the West to regions beyond—was slight. *Business Week* explained that in Arizona, Colorado, Idaho, Montana,

New Mexico, Utah, and Wyoming, railroads and trucking companies instead competed and intensely so, because industrial centers were few, far distant from one another, and freight moved on routes far distant from railroads.[156] According to the magazine, Arizona illustrated the region's typical condition. Seven licensed common carriers operated over three thousand miles of route, served all the principal towns, and shipped forty-two tons of freight daily. Three companies reported only small profits, and those that met railroad rates lost money.[157] In 1931, when subsidiaries were common elsewhere, railroads in the Southwest were trying to recover business lost to railroads, not business shared. Railroads in Colorado, New Mexico, Oklahoma, and Texas (as well as several states immediately east) joined together to file a tariff with the Interstate Commerce Commission to permit offering store-door and pickup service at a rate within three hundred miles of the shipments' origins (distance from big industrial centers), thereby gaining back 75 percent of the business lost to trucking.[158] At the end of 1939, *Railway Age* anticipated commencement "of the largest rail-highway operation ever attempted": the Southern Pacific's shipment of materials to railheads for truck distribution for the Colorado River aqueduct in southern California.[159] The trade journal reported it as consistent with railroading's increasing number of contracts, which had "put the railways into the motor transport business more largely and directly than ever before."[160] The West seemed to be growing consistently with the rest of the nation in this transportation matter.

Inter-Regional

The St. Louis Southwestern Railroad had a subsidiary, the Southwestern Transportation Company, which operated as far as Fort Worth, Dallas, and other eastern Texas towns by means of ownership, lease, and trackage rights, and benefited from considerable growth in business after the cessation of federal control during World War I.[161] Proportional to its size (service on 1,285 miles of highway), it operated the nation's largest highway freight line.[162] About a year after the company went into business, other railroads in the Southwest started subsidiary companies to operate intrastate in Texas and Louisiana. It seemed to *Railway Age* to have a promising future.[163]

Southern Pacific operated via two subsidiaries, Pacific Motor Transport Company and Pacific Motor Trucking Company, the first founded in 1929 and the latter in 1934.[164] The latter operated between Marshfield, Oregon, and, to the south, Tucson, Arizona, and the former within California.[165] In 1935 the Chicago, Burlington, and Quincy Railroad acquired a number of truck lines, after originally being limited only to bus lines, and operated through twelve states to the Pacific coast.[166] The truck subsidiary of the St. Louis & San Francisco Railroad, the Frisco Transportation Company, running through Arkansas, Florida, Missouri, Oklahoma, and Texas, by 1937 promoted its services to over a thousand stations in the South and Southwest with an LCL timetable in a booklet listing the stations serviced and in cartoons (fig. 9.8).[167]

Certain shippers using Union Pacific Stages, the Union Pacific Railroad's subsidiary, pushed the subsidiary to go beyond intrastate shipping in Kansas, Nebraska, and Oregon. At first business declined because of increased rates in response to the interstate market, but it improved, owing to the railroad's side of the bargain, by speeding up service and continuous demand for the interstate services.[168] This approach was novel. Most of the subsidiaries the trade literature remarked upon continued to be born of the railroad companies.[169]

FIGURE 9.8. A cartoon representative of the series to attract business to the Frisco Transportation Company depicted pride and prompt delivery. "Selling Railroad Freight Service," *Railway Age*, n.s., 107 (Nov. 25, 1939): 827.

Call and Delivery Service

PICK-UP and delivery is an outstanding feature of Railway Express Service in the cities and towns large enough to warrant it.

Express is complete through service. It functions, to a large degree, directly from the door of the shipper to the store or residence of consignee. A call by telephone, post, or in person, starts this service in operation.

When you have a shipment to make, you can depend on Railway Express. Over 70,000 experienced and alert railway expressmen eagerly await your call for service.

RAILWAY EXPRESS AGENCY, Inc.

Phone: 2-2117

510 CHURCH STREET
HARTFORD, CONN.

FIGURE 9.9. The Railway Express Agency advertised the ease and security on which its customers could rely. Railway Express Agency advertising flier, circa 1935. Authors' collection.

RAILWAY EXPRESS

The Railway Express Agency was founded in 1839 to carry small parcels, money, and important papers on trains, and a century later seventy of the nation's key railroads acquired its services, according to a 1939 article in *Railway Age*.[170] It ranked as one of the two largest motor truck operators in the country by 1930.[171] In 1932 the agency began work through a subsidiary, the Railway Express Motor Transport, Inc., and two routes were begun, one between Chicago and Milwaukee and another from Chicago to South Bend, Indiana (fig. 9.9). It marked the company's direct entry into highway transportation.[172] Two years later trucking reacted negatively against its railroad-owned competitor, as railroads had originally been very negative about trucking. However, the Interstate Commerce Commission dismissed the American Trucking Association's charge that the Railway Express Agency was unfairly conceived to circumvent the Interstate Commerce Commission and the Elkins Act of 1903, which fined transporters offering rebates.[173] In 1940, the agency operated about fourteen thousand trucks in the United States and Canada for pickup and delivery service to the railroads. This number of trucks ranked it second behind the United States government.[174] Operations were fully standardized across the nation.[175]

CONTAINERS

Business booming during World War I inspired equally creative ways to handle it physically in the transportation industry. Terminals could not keep pace, and congestion resulted. After the war, however, one solution was the "container car" for LCL delivery to the merchandise's destination, which facilitated handling and speeded up delivery. The first containers were truck bodies whose end and side doors permitted loading and unloading. The New York Central Railroad, renowned for various inventions that solved railroad-shipment problems, developed one car with nine containers to be loaded onto specially designed flatbed rail cars or an entire car fitted as a container. These saved the longer time otherwise taken to load and unload items individually. In a test run, it took twenty-seven minutes in New York to remove nine containers from trucks and twenty-one minutes in Chicago to offload them from the train onto trucks.[176] It cut back on the amount of labor required to box items one by one in individual containers required theretofore. The River and Rail Transportation Company of St. Louis simultaneously developed a similar, but not identical, system that was estimated to handle work at 300 to 400 percent less cost than the prevailing systems. Thus congestion, it was believed, would cease, and existing terminals would be entirely adequate if handling was expedited.[177]

Railroad executives ardently promoted the container cars from their inception.[178] "Freightainers" operated by the Boston & Maine Railroad varied insofar as a trucking subsidiary offloaded empty containers for the shipper to fill and then for the subsidiary to pick up when loaded; these specially constructed trucks with containers were for intercity transportation.[179] Indicative of their usefulness, container cars and shipping continued. They were cheaper than most LCL shipping because they eliminated handling, especially sorting, rehandling, and transfer delays; and they reduced loss and damage claims.[180]

Legal circumstances, however, stalled the use of container cars in 1929 and early 1930s. The Interstate Commerce Commission, in one instance, demanded a cost-study test to demonstrate that the cost of the service versus that of regular merchandise justified different rates.[181] That was in 1929. The ICC also forced the Chicago, North Shore & Milwaukee to abandon its policy of varying minimum container rates because only quality of service or cost of service could legally induce varying rates.[182] That was in 1931. Competing railroads and truckers had complained to the commission, and the decision ended "intermodal" transportation via containers for many years.[183] That was not the only hurdle for this innovation.

PIGGYBACK

Intermodal transportation's most consistently applied means was cargo shipped on flatcars, a practice known colloquially as "piggyback." Containers had always been curtailed to some degree because they required equipment not ready at hand, lying about in rail yards for utilization. Specialized equipment necessitated additional costs. When it first was used, piggyback received no notice in

INDUSTRY'S FOUL-WEATHER FRIEND

"In providing industry with a unique all-weather, door-to-door service, railroads have shown a determination to continue as the backbone of mass transportation."—Arthur K. Atkinson, President

FIGURE 9.10. Piggyback service depicted in winter on the Wabash Railroad. The line's president extolled a service he expected to grow "as it combines the inherent advantages of rail dependability and speed with the truck's door-to-door flexibility." Wabash Railroad timetable, October 27, 1957, inside back cover.

the trade literature. *Railway Age,* in 1950, did reflect back on the Chicago, North Shore, & Milwaukee Railroad's use of trucking firms' highway semi-trailers in 1926, its use two years later of the railroad's own semi-trailers for either those shipping or receiving, and in 1932 its use of loaded or empty semi-trailers by trucking firms. Service gradually increased because the railroad's rate was a little lower than it cost a trucking firm (fig. 9.10). The first region outside the Midwest or Northeast to use piggyback was the West, where the Denver & Rio Grande Western started piggyback service between Denver and Grand Junction, Colorado. Problems did beset the invention, however, including the requirement of specially equipped flat cars whose costs would have been less if the special cars could have been used when traffic slackened. Also, when fifteen or twenty cars were needed, lengthy loading times became an issue.[184]

TRUCKING REGULATION

As we reflect back on the two chapters necessary to cover trucking's relationship with railroads, it is noteworthy to recall that a professor of civil engineering speculated when he wrote in 1922 in *Highway Transportation* "that motor trucks are not only here to stay, but that their use will increase many fold."[185] With increased use came regulation; in 1924, *Railway Age* held that it was imminent. States should regulate intrastate trucking and the Interstate Commerce Commission trucking across state lines. Only thirty-one states had regulations regarding trucks. Amid several additional recommendations came the one that sparked the greatest disagreement over time regarding trucks and trains—that is, that trucks not drive out railroads.[186] What should trucking pay for the highways on which it depended for its livelihood? This question was implicit in the matter of threatening railroads.[187] *Railway Age* routinely reported the advancing cause of truck regulation.[188] By 1929, a milestone appeared to have been reached in the mind of the journal's editors: "It is notable in general that the regulation of motor transportation is becoming stabilized, similar to the way in which motor transportation is becoming stabilized."[189]

With the Great Depression, railroads cried out again against their diminished earnings and unfair competition. Truckers appreciated that their earnings should pay a part of the cost for highway construction and maintenance, *Road and Streets* editorialized in 1932. It pointed out, too, that only since the early 1920s had highway transportation taken traffic that railways might have gotten, and it was passenger traffic that lost the most.[190] The president of the New York, New Haven & Hartford Railroad spoke out for trucking to practice self-regulation rather than confront outraged public opinion.[191] California's experience with "wildcat" truckers in the early years of the Great Depression motivated the Railroad Commission of California to want the United States Supreme Court to take up issues that trucking, but especially wildcat truckers, brought to the fore. Failing resolution, at least for the wildcats, laws ought to prohibit their operation.[192] The Association of Railway Executives and the National Automobile Chamber of Commerce argued their ideas in 1932 before the National Transportation Committee (fig. 9.11).[193]

Eventually, the administration of Franklin Roosevelt, by means of the National Recovery Act, became the first to entangle the federal government in many of the questions.[194] In response to the National Recovery Act's stipulation for a "code of fair competition" from each industry, trucking's two organizations,

A Modern Gulliver

FIGURE 9.11. A cartoon portrayed the railroad's persistent claim of unfair competition from automobility's buses and trucks. "Equalizing Carrier Competition," *Railway Age*, n.s., 97 (Dec. 1, 1934): 699.

the American Highway Freight Association and the Federated Trucking Associations of America, agreed upon a code in 1933 and merged as the American Trucking Associations.[195] The Motor Carrier Act of 1935, discussed at greater length in the conclusion, made strides toward regulating trucking, but railroad executives remained disgruntled, convinced that trucking regulation had "so far been confined to the externals of the problem, such as the issuance of certificates of convenience and necessity and half-hearted attempts at rate regulation." That view was expressed by a Canadian railroad executive, whose speech made allusions here and there to the United States, including this quotation; the editor of *Railway Age,* which published the speech, assured readers that the conclusions "would probably apply with equal weight to conditions in the U.S.A."[196] In 1940 *Railway Age* doubted that the Motor Carrier Act of 1935 produced "a sound, coordinated national transportation system."[197] The year ended with passage of the Transportation Act of 1940, which added water carriers to all the commercial transportation means but put airlines under the regulation of the Interstate Commerce Commission.[198] Since the waning years of the nineteenth century, the more recently a transportation technology was used for commercial gain, the later it was placed under government regulation.

Trucks had come onto the transportation industry's scene even before their mechanical systems were as good as they could be. They entered very quickly. They were soon fundamental parts of the transportation system. When possible, railroads literally bought into their advantages, setting up subsidiary trucking companies. In their advancing wave of customer approval, railroads also quickly became uneasy with trucks. The transportation trade term "coordination" was what civic-minded people and many in the trucking and railroad industry both hoped for from the two systems. Contract carriers operating under licenses granted by governments were challenging enough to railroads. Where truckers operated independently, they seemed even more troublesome and all the more so when "gypsy" truckers were in business. Railroads wanted the latter disbanded and trucks to come under more regulation—for example, to pay more for highways. Railroads thus ardently favored government regulation that they had found unwelcome when applied to themselves, beginning in the late nineteenth century. Highway hauling continued to grow despite the Great Depression and the enactment of new regulations through the 1930s.

CONCLUSION

Automobility's rise was not something America's railroad executives could long ignore although many at first had done so. Rather quickly the motorcar, the motor bus, and the motor truck became innovations to be variously combated or embraced. Certainly government funding of highways needed to be discouraged. And bus and truck interests needed to pay their fair share of the taxes used to build those roads. They ought to be subject to equivalent regulation as commercial carriers of passengers and freight, and thus held to the same operating standards as the railroads. But railroad executives as well came to see motor technology as potentially helpful. For example, self-propelled rail motor cars were adopted but then, of course, discarded, advancing the idea of the streamlined train. Bus and truck subsidiaries were established, with the railroads, in fact, taking an important share of motor-carrier highway business.

The players were various, including not only railroad executives but also the financiers that sustained them, as well as politicians at all levels of government. There were the automakers (as well as the gasoline refiners, tire makers, and highway builders, among others) who parried railroad's complaints, lobbying strenuously in state legislatures and in Congress against reducing highway subsidies, raising taxes, and discouraging motoring in whatever manner. Consequently it was not motor vehicles that Americans rejected, but rather travel by streetcars, traction cars, and, to a large degree, passenger trains, especially the locals of the slow variety, including those with rail motor cars. Short-distance rail travel was substantially diminished in favor mainly of private automobiles, but also buses. Busing in cities became a kind of last resort for officials wanting to reduce the costs of public transit. Intercity busing became a refuge for those unable or unwilling to pay for faster train travel and ultimately even faster travel by air. Freight trains would be reduced to carrying mainly bulk materials, the railroad's package or less-than-carload business largely taken away by trucking.

Through the 1920s it had been assumed that railroading would always be central to fulfilling the nation's transportation needs. But instead the railroads had begun to falter. There was, however, a sense of mistrust implicit in the industry's great size, in what was perceived as its substantial influence over other economic sectors, and, of course, in its past misdeeds. Thus, increasingly severe regulation of the railroads had come to the fore, and just at the time that motorcars, trucks, and buses arrived. With its promise of speed, convenience, flexibility of action, and, importantly, self-gratification, motoring became increasing popular, especially with the arrival of mass-produced and, accordingly, relatively inexpensive automobiles. And commercial trucking and busing largely followed in its wake, given the new and improved highways that private motoring brought to the fore. Building a modern road system, although it was rationalized in Congress largely on the basis of providing "defense highways," was made politically feasible by widespread auto ownership and use driven by inexpensive, mass-produced cars introduced by the likes of Henry Ford.

In the 1930s, with the deepening of the Great Depression, partial remedies for railroad woes were attempted. First came the Emergency Railroad Transportation Act of 1933 and then the Motor Carrier Act of 1935. The earlier legislation sought to guarantee freight and passenger rates under purview of the Interstate Commerce Commission. Rates were to be adjusted and thus keep the nation's railroad's solvent. It tied railroad passenger and freight rates directly to railroad revenues, rates to be set just high enough for railroads to earn profits. One unanticipated consequence was that motor carriers were enabled to raise their rates also, resulting in a masking of many trucking and busing inefficiencies. The later act, which we emphasize in this conclusion, sought to level the transportation playing field by also placing bus and trucking companies under Interstate Commerce Commission regulation. Of course, what was missing, and is still largely missing today, was concern for integrated transportation planning.

For Congress the avoidance of monopoly in transportation remained primarily a matter of different transport modes competing with one another. Missing was the notion that various transport modes could be integrated within single corporations, with each mode contributing what it did best to an overall objective: providing the best transportation customers could afford. Winners and losers persisted instead, with the railroads tending to lose. Winners were the modes that continued to receive direct and indirect government subsidies: specifically the motor carriers but also the airlines and the inland-waterway barge interests. Having defined in the eighteenth century that harbor or port improvement, and the improvement of navigation generally, required federal support, the inland waterways came to benefit, even to the subsidization of a federally supported barge line. Similarly, the airline industry was subsidized with the creation of air "ports" and beacon "navigation systems" that enabled commercial air service to thrive. U.S. Mail contracts that enabled passenger air service to expand were also important.

The railroads, in contrast, had evolved through private enterprise, although federal subsidy had initially played an important role in the industry's creation, especially through the granting of federal land to sustain railroad construction, particularly in the Midwest and West. Nonetheless, by the 1880s the railroads had come to be viewed mainly as creatures of Wall Street and thus purveyors of financial greed, if not outright corruption. Farmers especially saw the railroads as a kind of enemy, one that in pursuing monopolistic ends from locality to locality manipulated freight rates to the farmers' disadvantage. Populist and then progressive political agendas emerged: the Sherman Antitrust Act was one result, and the Interstate Commerce Commission followed in its wake. The Sherman Antitrust Act was passed to reduce monopolistic railroad practices, if not railroad monopolies. ICC regulation would eventually restrict innovation in the railroad industry, mainly by restricting a railroad's ability to adjust rates, something necessary to meet the increasing challenges posed by rise of the nation's new automobility.

The Motor Carrier Act of 1935 established that bus and truck interests would be regulated much as the railroads were, an important step, it seemed, in finally leveling the nation's transportation playing field. Yet that was not fully the case. Was not Congress also acting to limit railroading's monopolistic potential? Was not the ghost of the Sherman Antitrust Act still implicit? Had not the railroads, through their bus and trucking subsidiaries, for example, come to hold a preeminent position in those industries? Let us then examine the 1935 act and its immediate aftermath. By doing so we bring our story up to World War II and the decidedly changed political and economic agendas that war dictated. It is with 1940 that we end our story. During the war, of course, railroad passenger service rebounded, a result both of the increased wartime travel demands and war-imposed gasoline rationing that vastly reduced private automobile use. Commercial trucking was also somewhat impaired, not only through gasoline shortages but also with new truck manufacture going mainly to meet war needs. Just as in World War I, the railroads found themselves confronting more passenger and freight haulage than could be efficiently handled. The railroad rebounded, but only temporarily.

THE MOTOR CARRIER ACT OF 1935

When the Transportation Act of 1920 was proposed, debated, and passed, the underlying assumption remained that "transportation" meant railroading. This assumption and the policies it put in place, especially tighter ICC regulation of railroad operations, quickly became outdated. By the end of the 1920s, not only had commercial busing and trucking matured but, again, so also had transport by barge on the nation's inland waterways; and an airline industry had been born. Increased use of private automobiles between 1921 and 1930 alone undercut rail passenger traffic by over 80 percent.[1] Although many observers had advocated the coordination of various modes of transport in integrated transportation companies, Congress persisted in seeing each mode as functionally separate. Competition between modes would remain the principal means by which competition in the transport field would be maintained. As the railroads had traditionally been predominant, surely they would remain so. With the railroad industry taking important ownership positions in busing and trucking, a continued recipe for continued monopolistic practice seemed well in place.

But the reality was that the railroad industry, with the nation's deepening economic downturn, was in serious difficulty. Accordingly, nationalizing the railroads was seriously discussed in Congress. During World War I, the railroads had been under federal control, but had been returned to private management by the Transportation Act of 1921. It was not the federal government's role, it was argued, to compete with private enterprise. That sentiment prevailed as the Herbert Hoover administration gave way to Franklin Roosevelt's New Deal. In 1932 stopgap actions by the Railroad Credit Corporation and the Reconstruction Finance Corporation, two Hoover administration initiatives, prevented a series of railroad bankruptcies. The Emergency Railroad Transportation Act of 1933 had followed. Then the 1935 Motor Carrier Act was passed. Ostensibly, its purpose was to prevent wasteful and destructive competition within the bus and trucking industries, with the Interstate Commerce Commission charged with exercising three principal oversight controls. First, entry into both commercial busing and trucking would henceforth require certificates of public convenience and necessity, which the ICC could deny. Second, bus and truck operators would be required to conform to regulations governing finance, accounting, record keeping, safety, and insurance, which, of course, the ICC would oversee. Third, carriers would be required to adhere to published rates and provide advance notice of changes, which the ICC was empowered to revoke.[2] Thus a Motor Carrier Bureau was set up within the ICC. The act did restrict somewhat railroading's motor-carrier initiatives. Neither the railroads nor their bus or truck subsidiaries could establish new routes or purchase established bus or truck companies if such actions diminished competition. Subsequent oversight by the Motor Carrier Bureau tended to restrict railroad subsidiary trucking primarily to localities previously served by rail. Pickup and delivery services closely linked to rail lines could flourish but not new long-distance trucking over routes beyond.

TRADE JOURNAL COMMENTARY

A sampling of opinion expressed in editorials and articles in trade journals of the period offers insight regarding the ways in which the 1935 Motor Carrier Act was anticipated, passed, and then assessed as to potential impact. The Interstate Commerce Commission opened hearings in St. Louis in 1930, its stated focus being the "coordination" of railroad and highway transportation. *Railway Age* commented on ICC chairman Ezra Brainer's opening comments. The railways, he said, "are the backbone of the transportation system of the country," but other forms of transportation, "particularly motor transport," had assumed "positions of importance." Some railroads, he noted, had made use of motor vehicles in connection with their rail operations but had encountered legal obstacles preventing the full coordination of rail and motor. What changes might be considered? Witnesses for various railroads and railroad motor-carrier subsidiaries then spoke. A representative from the Missouri Pacific Transportation Company asserted the need for motor-carrier regulation. Regarding busing: "There is no stability in rates, causing railways to make experimental reductions in railway passenger rates in an effort to meet the competition." The same is true, he said, of truck transportation. "All kinds of transportation should be subject to the same kind of regulation." The company's operations, "while unprofitable in themselves, have enabled the railway to effect substantial savings in train operation costs . . . making the highway operation profitable from the standpoint of the railway." And finally: "The bulk of the truck traffic moves by private or contract carriers . . . and regulation of only common carrier truck lines would not be sufficient to put truck and railway competition on a fair basis."[3]

The ICC fact investigation continued into 1931 as Thomas H. MacDonald, chief of the Bureau of Public Roads, came to testify. Thus *Railway Age* focused on "some rather sweeping statements" that MacDonald made regarding the gross weight of buses and trucks and "the little effect on the cost or the wear of a properly constructed highway." MacDonald had asserted that with "increased taxation in recent years," buses and trucks were "fully meeting all the excess cost of road construction made necessary by the increased loads." But under cross-examination he admitted that "the general standard of road construction is higher than would be required for more than 90 per cent of the traffic." Railroad interests had long argued that trucks were tearing up the nation's highways. Then MacDonald was questioned about the financing of road construction and the nature of highway traffic. Who paid for roads? And who used those roads? In 1929, he said, 81.9 percent of the money for state highway construction came from fees and gasoline taxes paid by the road users, including the 18.2 percent provided from bond issues, which were also almost wholly similarly financed. For the years 1929 and 1930, statistics collected in eleven western states suggested that not over 16 percent of the traffic on public highways was truck traffic, that not over 5.5 percent of that was common-carrier traffic, and that not over 8.7 percent was contract trucking.[4]

A representative of the Nashville, Chattanooga & St. Louis Railroad discussed (through "a series of exhibits of photographs, maps and statistical data") the history and cost of highway construction in Tennessee, the operation of motor vehicles on the state's highways, their effect on railway traffic, and the differences between ICC regulation of the railways and state regulation of buses and trucks. While the railroad paid out 33 percent of its earnings "for taxes and the maintenance of and interest upon its roadway," the trucks in the state were paying but 5 percent of their earnings for the same purpose, and the buses only 10 percent. He argued that truck and bus interests should be paying in addition to ordinary taxes "something in the nature of a rental for the use of public property as a place for doing business analogous to the rental paid by a railroad for the use of another's tracks." He said that as a private motorist he was paying the state eighty dollars a year in fees and taxes, "a large part of which represents a personal contribution to the expense of motor transport operation which is taking away a large part of the business of a railroad that was already providing adequate service."[5] At the same session a representative of the National Automobile Chamber of Commerce spoke against the railroads: "They continue to believe that regulation of the bus as an agency in interstate commerce is desirable in the public interest. They have not yet seen any facts which indicate that regulation of the truck as a common carrier in interstate commerce is practicable or that it is desirable in the public interest. Further they are emphatically opposed to any taxation or regulation of any motor vehicle which is undertaken solely for the purpose of 'equalizing competition.'" The chamber, in other words, opposed any action that might in any way reduce the sale of motorcars, buses, and trucks.

In 1933 Congress passed the Railroad Emergency Transportation Act. In so doing it appointed Joseph B. Eastman as federal coordinator of transportation, a post from which he was delegated to further investigate the railroad industry's financial stress, particularly in regard to the industry's relationships with other transportation modes. *Railway Age* interviewed Eastman early in 1934. Federal oversight of all forms of transportation was needed, he said, but with "little relaxation of the present degree of railroad regulation." Proposals submitted to his office by the Association of Railway Executives requesting a relaxation of railroad oversight had been "carefully analyzed," but most of the suggestions had been rejected. Yes, use by commercial buses and trucks' use in interstate commerce needed oversight. But he warned the railroads that "such regulation is in no way to be regarded as a panacea for railroad ills." The railroads, he said, "have spent too much time and attention on plans for the restriction of their competitors and too little time on the development and improvement of their own service and the readjustment of their own rates."[6]

As unemployment grew, the Roosevelt Administration sought stopgap measures to put workers back to work. Doing so would surely increase consumer spending and thus bring back a healthy marketplace. Since the mid-nineteenth century the railroad industry had been a significant consumer of manufactured goods and, indeed, one of the nation's biggest employers. But with the depression there had come not only a rapid decline of railroad revenues but also a rapid increase in the costs of equipment and materials needed in railroad maintenance and improvement. Purchases had been cut back drastically. Workers had been laid off. Articles appeared in every railroad trade journal arguing that healthy railroads were absolutely necessary to economic recovery, if only through their ability to purchase durable goods. American industry, it was argued, relied heavily on railroad purchases "for a normal volume of activity," and there was no hope to achieve economic rebound without "normal buying by the railroads."[7] Needed, among other changes, were substantive passenger- and freight-rate increases.

But action in Congress on new transportation legislation seemed stalled. Then the Senate passed a bill and sent it to the House, which rejected it on the grounds that it had too limited an application. It applied only to steam and electric railroads (and related express companies and sleeping-car operators) and to pipelines and steamship lines. Highway motor carriers would continue free of federal regulation. Rejection reflected strong opposition from farm interests, which saw in the bill a threat to the private trucking of farm produce. The editors of *Railway Age* editorialized: "The present railroad situation and the neglect of it in Washington strikingly illustrates that in a democratic country politics is regarded by most politicians as more important than economics, and that when they are confronted with political issues apparently involving their own future, and economic issues involving the future of the nation, most of them give complete priority to the political issues."

But various other concerns were sidetracking Congress. There was, for example, the veterans' bonus issue that culminated in 1932 with the Bonus March on Washington, D.C. As *Railway Age* noted, there were "about 1,500,000 persons unemployed on the railways and in the durable goods industries because of the inadequacies of railway gross and net earnings." That was more than the total numbers of American soldiers that saw service in actual combat in World War I, and one-seventh of the total number of persons unemployed in the country. "And yet consideration of transportation and all other measures that promote recovery has been delayed for weeks by purely political maneuvering regarding veteran's bonus legislation."[8]

Finally, in August 1935 a comprehensive transportation bill was passed with motor-carrier regulation as its center piece. "Ten-year campaign for regulation ended when Senate accepted House amendments," headlined an editorial in *Railway Age*. Passage of the bill had been supported "not only by the railroads, but by the state commissions and the principal organizations of truck and bus operators. . . . It was opposed to the last by the National Highway User's Conference and by organizations representing the farmers." Joseph Eastman was quoted as saying, "I think it will help the railroads and also help the bus and truck industries, and will not eliminate competition in the transportation field. . . . The chief beneficial effect it will have will be in stabilizing conditions by preventing demoralizing and destructive competition. It ought to lay the foundation for sound future development in transportation and bring about better co-operation and co-ordination in that industry."[9] A new Motor Carrier Division was to be set up in the Interstate Commerce Commission, but then Congress adjourned for its summer break without passing a revenue bill authorizing it. Thus the original bill's impact was delayed. But *Railway Age* could not complain. Congress, strongly influenced by the Roosevelt administration, had passed an important bill potentially favorable to railroading. The administration had "shown more disposition to promote legislation especially to help the railways than any other administration for a quarter century."[10]

But from one month to another, *Railway Age* editors did very much question New Deal initiatives. And guest commentators did so as well. Fitzgerald Hall, president of the Nashville, Chattanooga & St. Louis Railroad, harped on the ICC's continued "over-regulation" of the railroad industry. Additionally, Franklin Roosevelt, he said, was in the process of distorting "republican institutions into a paternalistic all-powerful, centralized government." Thus, he argued, all citizens had cause for concern: "No industry in America has suffered from political regulation as have the railroads, and their plight should be a warning of what business generally may reasonably anticipate should this administration be successful in imposing upon the American people its fantastic measures designed to subject us all to bureaucratic regimentation from Washington." He reiterated: (1) Did not the federal government still restrict railroad innovation? (While the railroads under federal law might abandon neither old nor build new lines without the approval of a federal bureau in Washington, there was no similar inhibition upon any other form of transportation.) (2) Was there yet a "level playing field"? ("Of the larger forms of transportation in this country, the airlines, the boat companies on the inland waterways and the motor vehicles are given constant and direct financial aid through governmental subsidies. Only the railroads are expected to furnish and maintain their own facilities and pay all the operating costs.")[11] It would be some forty years before another president, Ronald Reagan, thinking the same kinds of thoughts, would push through Congress legislation substantially erasing railroading's crippling federal oversight and, as well, much of the oversight extended in 1935 over interstate busing and trucking. It would be four decades before the Interstate Commerce Commission would be eliminated, to be replaced by a less invasive Surface Transportation Board.

Soon after passage of the Motor Carrier Act, *Railway Age* began another editorial initiative: renewing the idea of integrated, multimode transportation companies, something the Eastman-led ICC sleuthing had not fully considered. Indeed, it might be said of Joseph Eastman's investigations as federal coordinator of transportation that he probably did more thinking about sustaining ICC bureaucratic reach than he did about solving the nation's transportation problems. Potentially, an integrated transportation company would act to amplify transport efficiency through intermodal service. It would, in the words of the journal, "seek out the cheapest means or combination of means of moving traffic and would not be interested in having moved most of the way by truck, railroad, or water, as the case might be, because of any special interest in some peculiar form of transportation. . . . The establishment of transportation companies would of course mean the abandonment of competition between the various types of transportation agency as a means of regulating rates."[12]

Trade journals serving the bus and truck industries were of more recent origin and thus less well established than *Railroad Age,* the railroad industry's dominant magazine. *Motor Truck News* was only two decades old, while *Bus Transportation* had been publishing for less than a decade. *Railway Age* had roots going as far back as the 1850s. The focus in bus and truck journals tended more to reporting corporate news. There was less coverage of (and, perhaps, less need to be preoccupied with) government regulation. But always those journals had rebutted the railroads' insinuations that buses and trucks operated through government subsidy. They rebutted the notion that motor vehicles were not paying their fair share of taxes supportive of the nation's new highways. "Special Motor Vehicle Tax Levies Top State and County Road Costs" was, for example, the headline of a 1938 article in *Motor Truck News.* The motorist driving along the highway and the freight truck loaded with goods for market were paying for the road as surely as though a meter were "clicking off nickels and miles." Private cars paid $38.78 a year on average in special motor taxes. The common carrier truck paid more than six times that amount "for the use of the road." All trucks, including light delivery trucks and farmers' vehicles, averaged two and a half times as much. The heaviest contributor, however, was the motor bus, which paid twenty times the amount of the private car. All told, motor vehicle taxes, including gasoline taxes, totaled some $1.3 billion annually.[13]

The "Motor Transport Section" in *Railway Age* was enlarged year by year. In it the magazine's editors continued to aim accusations at bus and truck operators. They were also concerned to know whether the new Motor Carrier Act was or was not working to railroading's advantage, dutifully reporting on decisions made by the ICC. For example, in 1936 the Motor Transport Section had vetoed the Pennsylvania Railroad's proposal to buy several trucking companies. Purchases were to be financed by the American Contract & Trust Company, a wholly-owned subsidiary of the railroad. The bureau objected on the grounds that such "self-financing" was untenable.[14] No competing railroad or trucking company had filed a complaint. Nor had any governmental entity. What was the ICC's objection? "The railroad-controlled finance company was not under *direct* oversight of the ICC," *Railway Age* noted, "and thus it could not approve or disapprove any proposal advanced with its involvement."[15] However, a year later, with changed financing, the railroad was able to complete its purchases. Another early test case involved the Pan-American Bus Lines, which had begun bus service connecting New York City with Miami. In this instance formal protest did come from literally every railroad and every bus company whose territory the initiative impacted. The service was ruled unnecessary since there were already sixty scheduled motor-bus runs between the two cities daily.[16]

SCHOLARLY COMMENTARY

The Motor Carrier Act of 1935 drew scholarly attention immediately.[17] James C. Nelson, an economist who had just finished a PhD dissertation on the history of motor transport regulation in the United States, published an article on the act in the *Journal of Political Economy* in 1936.[18] Overlooking agriculture's dissent, the act passed, he claimed, because the general public lacked concern. The only strong objections, he opined, had come from "the automobile manufacturers, who believed that the purpose of the I.C.C. in regulating motor carriers would be to restrict the growth of motor transportation for the benefit of the railroads, thus diminishing the demand for trucks and buses." Some shippers had objected. "This apprehension was based upon an observed tendency for state regulatory agencies to relate truck rates to the higher rail basis." Some contract truckers, he noted, as well as a few common-carrier truckers, feared that regulation would destroy the flexibility of their service, their greatest competitive advantage over rail carriers, and that rate regulation, if effective, would destroy the advantage they enjoyed because of their freedom to undercut rail rates at will. But public indifference was prime. It was part of the "usual apathy toward public matters." It marked a failure of most Americans "to comprehend the long-run effects of an oversupply of transport service" and partly by their "vague general belief [that] the consumer gained from low competitive transport rates." Finally, he argued, "Considering the many conflicting opinions and the generous misapplication of fact used in propaganda issued by partisan bodies, the indifference of the general public cannot be a subject of wonder."[19]

In 1940 economist Julius H. Parmelee's book *The Modern Railway* appeared. In it he dedicated one chapter to national transportation policy in general and another chapter to the Motor Carrier Act of 1935 specifically. Regarding general policy, he stressed the role played by the ad hoc National Transportation Committee of 1932, which had deliberated at the end of the Hoover administration.

The committee had been chaired by former president Calvin Coolidge, with Bernard Baruch as vice chairman and former New York governor and presidential candidate Alfred E. Smith, among other notables, serving. The panel advocated "standardized government regulation of all transport modes without due interference, with complete equality of opportunity, and without subsidy or other assistance from the government." In particular, the committee advised: "water, motor, and air transport should pay their own way."[20]

The forty-page act, Parmelee noted, was divided into two sections. The first was the original Interstate Commerce Act (as amended over the years), which, of course, applied only to the railroads. The second spelled out specifically how motor carriers were to be regulated also. Thus, together the two parts seemingly spelled out "a single comprehensive piece of legislation." But did it? Confusion, for example, arose relating to how motor carriers should be categorized. Various kinds of bus operation were clearly exempted from the act, including the transporting of school children. Taxicab operations were as well. But some activities seemed to fall between the cracks. There was the case of Scott Brothers, Inc., which performed freight collection and delivery courier services in and around New York City, particularly for the Pennsylvania and Long Island railroads. The question arose: did that company have to apply for a certificate as a contract carrier under the Motor Carrier Act? The ICC held that the company did, but since it was "performing a branch of railway service" its operations fell under Part I and not Part II. Such were the consternations fostered by the act's wording in its early years.

Parmelee emphasized that many problems in the field of motor-carrier regulation still awaited solution. How was federal subsidy of highway construction to be factored in? And how were the very small trucking operations to be treated? In 1940 most trucking operations were no more than single-person and single-vehicle operations. And the ICC had yet to bring the various transportation modes fully under uniform regulation. But, as Parmelee saw it, "the Interstate Commerce Commission was attacking problems one by one," although proceeding cautiously but steadily "to occupy the field outlined in the act, and will eventually have the whole field under control." Only then would it be possible to appreciate the soundness of the legislation regarding its long-term economic effect.

Of course, World War II intervened to create a wholly new economic circumstance. During the war private automobile use declined, and busing and trucking were placed on a plateau. But after the war automobility exploded, with the motorcar becoming even more a necessity and not just a luxury. Long-distance bus service thrived but mainly in service to less affluent Americans who still did not own cars or were not enamored of more expensive forms of transport. Truck transportation, however, took off, as truckers fully replaced the railroads in the carrying of package or less-than-carload freight. Today the railroads thrive mainly through bulk commodity commerce and the carrying of heavy merchandise, including, interestingly enough, automobiles. However, the Motor Carrier Act of 1935, in retrospect, did enable the railroads to achieve a degree of parity with motor carriers. It did excite a kind of truce between vested interests. Of course, the fact that the railroads were still very much involved in busing and trucking helped in that regard.

AMERICA'S RAILROADS ON THE THRESHOLD OF WORLD WAR II

Back in 1934 Joseph Eastman published an article in *Bus Transportation* in which he wrote:

> Before the automobile swept on the scene like a tropical hurricane, the railroads had very little highway competition. They had water competition here and there, and some pipeline competition, and they competed with each other, but the highways were feeders and not competitors. The railroads were quite content in those days to have the roads maintained by general taxation. Suddenly a vast system of new or improved highways was constructed over which swarms of new vehicles operated which offered throughout the nation competition for both freight and passenger traffic which had hitherto been deemed invulnerable to outside competition.

But to make matters worse for the railroads, "in reliance on this supposed invulnerability, and with general approval as well, they had biased their rates on cost of service only in part and had favored the lower-valued commodities and the longer hauls at the expense of the higher-valued commodities and the shorter hauls." This system of rates was "made to order for the new competitors," he wrote.[21] Federal regulators allowed railroad rate adjustments to guarantee minimal corporate profitability. In the process they also sustained competing transport modes. In the words of economists Robert Gallamore and John Meyer: "Railroad rates were held high as an 'umbrella' protecting less efficient (but nominally lower-priced) competitive modes."[22] No better summary could have been written. It brought the heart of the conflict between railroading and busing and trucking into focus.

Looking toward passage of the Motor Carrier Act the next year, Eastman wrote: "The railroads have always been classed as public highways and transportation in general as a public business. The government must see to it that this public business is preserved from disorder and waste and that it furnishes good service to all on equal terms and at known reasonable rates."[23]

In 1940 the steam railroads were indeed still primary in the nation's transportation scheme of things. But it was a declining primacy. They dominated (along with shipping on the Great Lakes and on the inland rivers) the long-distance hauling of bulk commodities, but package and less-than-carload freight, even over long distances, was slipping away rapidly to trucks. The automobile had taken much of the railroad passenger traffic, first for short-distance travel and then even over long distances. And commercial busing had fully matured both for intracity and intercity travel. The electric interurbans were all but gone. Streetcars were barely holding on. In 1937 ton-miles of freight carried by various transportation modes broke down as follows: (1) steam railroads, 64.6 percent; (2) Great Lakes shipping, 16.6 percent; (3) pipelines, 8.0 percent; (4) motor carriers, 7.0 percent; (5) inland waterways, 3.0 percent; and (6) electric railroads, 0.1 percent. Passenger traffic carried by various commercial transport modes broke down like this: (1) steam rail roads, 52.5 percent; (2) motor carriers, 42.7 percent; (3) waterways, 2.8 percent; (4) electric railroads, 2.0 percent; and (5) air carriers, 1.0 percent.[24]

In 1900 there were but eight motorcars registered in the United States.[25] In 1940 there were approximately 27.4 million, along with some 101,000 motor buses and 4,900 motor trucks.[26] Surfaced rural road mileage stood at 1.3 million miles.[27] There were 44,000 locomotives operating on American railroads, and 2 million freight cars and 45,000 passenger cars.[28] Steam railroad mileage still stood at 249,000 miles.[29] The survival of rail lines, however, had as much to do with the ICC preventing line abandonment than it did with railroad viability. Since 1900 changes in the nation's transportation system had been immense. Above all, the railroad industry had slowly lost control of its destiny, first to government oversight and then to new modes of transportation that largely avoided government regulation and even taxation. They had also benefited, as the railroads had not, from substantive government subsidies for building physical infrastructure. The major agent of change was the nation's ever-increasing automobility. But was motoring yet paying its fair share?

The railroad industry's deteriorating economic situation, exasperated by the Great Depression, was complicated not only by the public's preference for motoring and the rise of commercial busing and trucking, but additionally by the physical demands that streets and highways had placed on railroad infrastructure. The railroads were challenged to physically accommodate new automobile technology, especially at crossings where railroad trains and motor vehicles were placed in potential physical contact. Steam railroads found themselves helping to finance crossing improvements, including bridges and viaducts intended to separate rail from motor traffic. They found themselves fully financing elevated rights-of-way and even new bypass lines to help avoid city street congestion. Electric interurban lines found their effectiveness undermined by increased motor traffic on the streets and roads where they operated. Traffic congestion put an end to many a city streetcar line, with traction service being replaced by motor buses and electric buses. In the cities truck-convenient freight terminals were required. And passenger terminals, if not replaced, needed modification, making them more convenient for taxis and private cars. Parking became a necessity.

Not only did the railroads adapt, but they were also required to adopt. Automobile technology was variously turned to railroad purposes; motorized track maintenance and repair vehicles, for example, were provided with internal combustion engines. Important were the self-propelled rail motor cars that replaced steam trains, which were considerably more expensive to operate. Their perfection led to modern streamlined passenger trains, which did help in the 1930s to sustain long-distance railroad passenger service. Innovation on the railroads, in turn, led to technology transfers back again, as in, for example, the use of diesel engines adapted to buses and trucks after World War II.

That America's railroads had declined and could no longer claim to be overwhelmingly the nation's primary transportation mode was becoming clear by World War II. Locomotives—especially those that were steam-driven, with long trains of cars behind—still sped through cities and countryside with whistles and bells commanding attention. Such spectacles once sustained railroading as an essential element of American life. But if railroading still impressed the public after the war, it was more likely the new diesel klaxons and the shining new streamliners that promised a degree of renewal. But there were other machines at which to marvel, not the least of which were the new cars, trucks, and buses, if not the new passenger planes descended from those that had proven so vital in winning the war. That railroading might slip back into decline was a little alarming. But, nonetheless, new modes of transport did speak of overall progress, especially with the new interstate system of freeways proposed, discussed, and then launched in the Dwight Eisenhower administration.

Had automobility finally superseded railroading as a form of land transport? Automobile ownership and use had (or seemingly had) empowered the individual through enhanced geographical mobility, which, in turn, tended to enhance social mobility as well. The automobile had become a status symbol with pride of ownership clearly implicit. But for most Americans it had also become a necessity as landscapes and places had been largely reorganized around automobile accessibility. Motoring convenience had garnered popular political support for new highways, which then fostered not just more private motoring but the rise of commercial trucking and busing. Automobility had spawned public policies about which people in the railroad industry could only complain.

CODA

In writing previous books concerned with the automobile and its impact on American society and culture, we emphasized the "built" environment, especially the rise of "Roadside America," with its garages, gas stations, motels, fast food restaurants, advertising signs, parking lots, and so on.[30] Herein we have chosen to emphasize how automobilty, defined as the utilization of motor vehicles, overcame railroading as a preferred means of transportation. We have placed emphasis more on the process of change than on accompanying material culture. But the coming of motorcars, buses, and trucks wrought great change in railroad landscapes while creating new landscapes of their own. Substantially, the change was that of subtraction rather than addition. It was not so much what was built anew, but what was chosen to survive and what was not. With deteriorating revenues and, perhaps more important, facing ever-increasing tax obligations on physical plant, railroads met the future not by abandoning the past but by destroying much of its physical evidence. Unprofitable lines, although abandoned with difficulty, given ICC oversight, were, nonetheless, abandoned. Unused buildings were demolished. Obsolete rolling stock was sent to scrap yards. Thus relatively little remains of early-twentieth-century railroading's landscapes and places. Historic buildings, especially old depots, are seen here and there, but places made up of historic building ensembles are largely gone.

Rights-of-way may survive, but their historical furnishings are missing. If old railroad rolling stock exists, it does so mainly in museums. Between 1900 and 1940 both railway and highway corridors saw constant change. Although much of the early automobile era is now gone also, more, in relative terms, survives. Automobile-impacted places have changed more through accretion, the obsolete shoved aside by the new but replaced by something. Much railroad-era infrastructure, on the other hand, has been summarily destroyed and replaced with little if anything at all—railroad spaces and places stripped down to their bare essentials.[31] But all is not lost.

NOTES

Chapter 1

1. Slason Thompson, *Cost Capitalization and Estimated Value of American Railways: An Analysis of Current Facilities* (Chicago: Gunthorp-Warren, 1908), 46.

2. George R. Martin, "Reasons for the Railway Land Grants," *Railway Age*, n.s., 92 (January 30, 1932): 255.

3. *Statistical Abstracts of the United States, 1930* (Washington, DC: United States Department of Commerce, 1930), 393.

4. Ibid., 397.

5. "Our Railroad Development," *Literary Digest* 26 (June 27, 1903): 917.

6. Don L. Hofsommer, "Edwin Hawley: 'The Little Harriman,'" *Railroad History*, no. 2101 (Spring-Summer 2014): 68.

7. B. F. Yoakum, "The High Cost of Railroading, the Biggest of Our Industries," *World's Work* 24 (Oct. 1912): 649.

8. I. Leo Sharfman, *The American Railroad Problem: A Study in War and Reconstruction* (New York: Century, 1921), 38.

9. *Smyth v. Ames,* 169 U.S. 466, quoted in Julius H. Parmelee, *The Modern Railway* (New York: Longmans, Green, 1940), 351.

10. *U.S. v. Trans-Missouri Freight Association,* 166 U.S. 290. See Emory R. Johnson, "The Courts and Railway Regulation," in *American Railway Transportation* (New York: D. Appleton, 1914), 386–407.

11. Frank N. Wilner, *Railroad Mergers: History, Analysis, Insight* (Omaha, NE: Simmons-Boardman Books, 1997), 13.

12. See Harold U. Faulkner, *The Decline of Laissez-Faire,* vol. 2 of *The Economic History of the United States* (New York: Rinehart, 1945), 191–219; and Wilner, *Railroad Mergers,* 12–19.

13. *Northern Securities Co. v. United States,* 193 U.S. 197.

14. Yoakum, "High Cost of Railroading," 657.

15. "A Decade of Railroad Marvels," *World's Work* 21 (January 1911): 13830.

16. French Strother, "A New Day for the Railroads," *World's Work* 40 (June 1920): 197.

17. Johnson, *American Railway Transportation,* 418.

18. Roy V. Wright, "Our Railroads Under Government Control," *World's Work* 36 (July 1918): 293.

19. George Soule, *Prosperity Decade: From War to Depression, 1917–1929,* vol. 8 of *The Economic History of the United States* (New York: Rinehart, 1945), 35.

20. Sharfman, *American Railroad Problem,* 167.

21. Wilner, *Railroad Mergers,* 32, 33.

22. Pierce H. Fulton, quoted in "What's the Matter with Railroad Consolidation?" *American Review of Reviews* 77 (Feb. 1928): 221.

23. "Control of Railroads Analyzed for House Committee," *Railway Age*, n.s., 90 (Feb. 28, 1931): 450, 453.

24. "What About Railroad Stocks?" *World's Work* 48 (July 1924): 332.

25. James J. Hill, "After Effects of the War on Business and Railroads," *Trade and Transportation* 15 (Dec. 1914): 8.

26. Ivy L. Lee, "Burdens from Which Railroads Suffer," *Trade and Transportation* 14 (Feb. 1914): 14.

27. "The Railroads and Investors," *World's Work* 32 (June 1916): 142.

28. L. E. Johnson, "Some Aspects of Government Ownership," *Trade and Transportation* 16 (Dec. 1915): 10–11.

29. Aaron Austin Godfrey, *Government Operation of the Railroads: Its Necessity, Success, and Consequences, 1919–1920* (Austin, TX: San Felipe Press, 1974), 5–6.

30. Mark H. Rose, Bruce E. Seely, and Paul F. Barrett, *The Best Transportation System in the World: Railroads, Trucks, Airlines, and American Public Policy in the Twentieth Century* (Columbus: Ohio State University Press, 2004), 30.

31. Wilner, *Railroad Mergers*, 92.

32. Julius H. Parmelee, "A Review of Railway Operations in 1932," *Railway Age*, n.s., 94 (Feb. 4, 1933), 137.

33. "The Decline in Passenger Traffic," *Railway Age*, n.s., 86 (Feb. 16, 1929), 397–98.

34. Samuel O. Dunn, "The Future of the Railways," *Railway Age*, n.s., 95 (Oct. 21, 1933), 555–56.

35. "Railway Trends in 1932 and Prospects for 1933," *Railway Age*, n.s., 94 (Jan. 26, 1933), 130.

36. George E. Boyd, "Railway Construction Still Recedes," *Railway Age*, n.s., 81 (Jan. 4, 1936): 54.

37. J. G. Lyne, "Bankruptcies Break Record in 1935," *Railway Age*, n.s., 102 (Jan. 4, 1936): 41.

38. Eliot Janeway, "What Ruined the Railroads?" *Nation* 145 (Nov. 20, 1937): 556.

39. William Z. Ripley, "A New Step with the ICC," *World's Work* 60 (Dec. 1931): 20.

40. William R. Childs, *Trucking and the Public Interest: The Emergence of Federal Legislation 1914–1940* (Knoxville: University of Tennessee Press, 1985), 119.

41. Wilner, *Railroad Mergers, 72.*

Chapter 2

1. H. W. Howard, "The Development of Highway Transportation," *Highway Transportation* 13 (Oct. 1923): 5.

2. *Automobile Facts and Figures*, 21st ed. (Detroit: Automobile Manufacturers Association, [1939]), 4. This volume provides figures for calendar year 1938.

3. "Examinations for Automobile Operators," *Railroad Gazette* 32 (1900): 125.

4. "Automobile Registration," *Street Railway Journal* 22 (Aug. 22, 1903): 249.

5. "New York's Auto Exports Increase," *New York Times*, July 14, 1913, 11.

6. "The Automobile and the Law," *Street Railway Journal* (Aug. 4, 1904): 175.

7. *Railway Age*, n.s., 40 (Oct. 20, 1905): 477.

8. "Automobile Versus Railway," *Literary Digest*, 27 (Sept. 19, 1903): 354.

9. "The Influence of Automobiles on Railways," *Railway Age*, n.s., 37 (Jan. 15, 1904): 107.

10. "Automobile Races for Vanderbilt Cup," *Railway Age*, n.s., 38 (Oct. 14, 1904): 559; *Railway Age*, n.s., 40 (Oct. 20, 1905): 477.

11. For example, see "Automobile Racing," *Railway Age*, n.s., 41 (Jan. 26, 1906): 163; "New Transcontinental Automobile Record," *Railroad Gazette* 41 (Aug. 17, 1906): 42; and *Railroad Gazette* 51 (Feb. 2, 1906): 101.

12. *Frank Leslie's Popular Monthly*, cited in "Progress of the Automobile," *Literary Digest* 28 (1904): 175.

13. "Progress of the Automobile," 176.

14. "What the Automobile Is Doing," *World's Work* 11 (Feb. 1906): 7262, 7263.

15. M. C. Krarup, "Automobiles for Every Use," *World's Work* 13 (Nov. 1906): 8163–78.

16. Rudi Volti, "A Century of Automobility," *Technology and Culture* 37, no. 4 (1996): 667.

17. "The Automobile in Railroad Rivalry," *Railroad Age Gazette* 45 (Dec. 4, 1908): 1465–66.

18. "The Railway and the Automobile," *Railway Age Gazette* 49 (Dec. 9, 1910): 1103.

19. W. R. McKeen, "The Value of Motor Cars," *Railway Age Gazette* 59 (Oct. 29, 1915): 817.

20. Merle Shepard, "Will Build 1,250,000 Cars in 1916, with a Valuation of a Billion Dollars," *Spokesman*, Mar. 1916, 158, 160.

21. "Automobiles as Preventatives of Railroad Building," *Railway Age Gazette* 6 (Oct. 13, 1916): 1.

22. "How Fords Have Multiplied," *Auto Dealer And Repairer* 24 (Oct. 1917): 41.

23. "Motorcar Travel Greater Than Steam-Railroad Travel," *Literary Digest* 54 (Jan. 20, 1917): 165–66.

24. "Agricultural States Leaders in Cars and Trucks," *Automotive Industries* 39 (Aug. 22, 1918): 308.

25. Samuel M. Felton, "New Influences Affecting Passenger Traffic," *Railway Age Gazette* 61 (Dec. 1, 1916): 997.

26. "Six Billion Miles of Motor Trucking," *Federal Traffic News* 6 (July 15, 1919): 2.

27. "A Year of Prosperity for Railway Labor," *Railway Age Gazette* 66 (Jan. 3, 1919): 37–42.

28. "How War Hit Auto Industry," *Highway Transportation* 8 (May 1919): 19.

29. "6,353,233 Cars and Trucks in Use in United States," *Automotive Industries* 41 (Aug. 28, 1919): 402.

30. "Automobile Transportation Costing as Much as Railroad Transportation," *Railway Age* 69, no. 6 (Aug. 6, 1920): 214.

31. "Red Automobile Tail-Lights," *Railway Age*, n.s., 72 (Jan. 14, 1922): 166.

32. "It Is Getting Worse Every Minute," *Railway Age*, n.s., 74 (May 12, 1923): 1143.

33. "What the 'Automobile Guest' Does to the Electric Railway," *Electric Railway Journal* 64 (Nov. 29, 1924): 907–8.

34. C. M. Burt, "Effect of Automobile Travel," *Railway Age*, n.s., 77 (Oct. 11, 1924): 643.

35. "Railroads and Private Motorists," *Railway Age*, n.s., 93 (Sept. 3, 1932): 317.

36. Harland Bartholomew, "Decentralization a Real Answer to Street Congestion," *Electric Railway Journal* 63 (Mar. 8, 1924): 366.

37. "Railway Passenger Travel," *Railway Age*, n.s., 70 (July 4, 1925): 2.

38. "Boston & Maine Allowed to Abandon 58 Miles of Line," *Railway Age*, n.s., 79 (Nov. 28, 1925): 1002.

39. "Causes of Railroad Abandonments," *Railway Age*, n.s., 79 (Nov. 7, 1925): 865.

40. Elisha Lee, "The Relation of the Railroad and Motor Vehicle," *Highway Transportation* 12 (Jan. 1923): 13.

41. Ibid., 29.

42. "Bus and Truck Operation Should Be Co-ordinated," *Railway Age*, n.s., 80 (Jan. 16, 1926): 231–33.

43. Henry R. Trumbower, "Highways Largest Passenger Carriers—Railways, Freight," *Railway Age*, n.s., 71 (Sept. 25, 1926): 608–9.

44. *Highway Motor Transportation: Report of the Subcommittee on Motor Transportation of the Committee for the Study of Transportation* (N.p.: Association of American Railroads, Aug. 1945), xxviii.

45. "Getting Automobile Owners as Railroad Passengers," *Railway Age*, n.s., 71 (May 22, 1926): 1360.

46. General W. W. Atterbury, "Looking Ahead in Transportation," *American Review of Reviews* 79 (Apr. 1929): 59–62.

47. "3,634,272 Motor Vehicles Are in Use Outside United States. World Total Now Is 31,360,779," *Automotive Industries* 52 (Feb. 26, 1925): 343.

48. "The Conquering Automobile," *World's Work* 52 (Dec. 1928): 124–25.

49. Samuel Crowther, "Automobiles: Carriers of Progress," *World's Work* 59 (May 1930): 39–41.

50. *Facts and Figures of the Automobile Industry*, 1931 ed. (New York: National Automobile Chamber of Commerce, 1931), 10, 86.

51. "Preliminary Facts and Figures of the Automobile Industry in 1935," *Motor Age* 55 (Jan. 1936): 47.

52. *Highway Motor Transportation*, 171.

53. Ibid., xxvii.

54. Robert C. Lieb, *Transportation: Domestic System* (Reston, VA: Reston Publishing, 1978): 6, 72–74.

55. George H. Douglas, *All Aboard!* (New York: Paragon House, 1992): 323–27.

56. John A. Heitmann, *The Automobile and American Life* (Jefferson, NC: McFarland, 2009): 206–7.

57. Douglas, *All Aboard*, 317.

58. Isaac B. Potter, "History of the Movement," *Good Roads Magazine* 33 (May 1903): 168–70.

59. George L. McCarthy, "How Railroads Are Interested," *Good Roads Magazine* 32 (Dec. 1901): 6.

60. *Railway Age*, n.s., 32 (Nov. 1, 1901): 487.

61. "Good Roads Convention," *Railway Age*, n.s., 35 (Feb. 27, 1903): 306; "A Maximum of Good Results: Martin Dodge and the Good Roads Trains," Highway History, Federal Highway Administration website, http://www.fhwa.dot.gov/highwayhistory/dodge/09.cfm (accessed July 22, 2014).

62. "Railroad Aiding Road Construction," *Good Roads Magazine* 49 (Jan. 8, 1916): 29.

63. "Road Building During The War," *Good Roads Magazine* 51 (June 2, 1917): 329.

64. "Main Highways Are of Real Military Importance," *Highway Transportation* 8 (Dec. 1918): 9–10.

65. "State Road Building and Railroads," *Railway Age,* n.s., 68 (Mar. 16, 1920): 810.

66. "Supreme Court Sustains Railroad in Objection to Assessment for Highway Construction," *Railway Age,* n.s., 70 (June 17, 1921): 1420.

67. "Railway Assessments for Permanent Highway Construction," *Railway Age,* n.s., 70 (June 24, 1921): 1428–29.

68. "The Railways' Interest in Highway Construction," *Railway Age,* n.s., 71: 7 (Aug. 13, 1921): 278.

69. C. E. Jones, "Should the Government Build Railways or Highways?" *Good Roads Magazine* 53 (Mar. 23, 1918): 171.

70. R. E. Fulton, "Good Roads Situation Is Like Railroad Problem 50 Years Ago," *Highway Transportation* 8 (June 1919): 35–36.

71. "Highways and Railways," *Highway Transportation* 9 (Mar. 1920): 21.

72. "Reconstruction of Narrow Roadways of Trunk Highways," *Highway Transportation* 9 (Jan. 1920): 17, 45; Arthur H. Blanchard, "Paved Highways for Present Day Traffic," *Highway Transportation* 11 (Dec. 1921): 1212–13, 1238.

73. "The Public Should Be Told the Facts," *Railway Age,* n.s., 72 (1922): 1222.

74. Frank Terrance, "The Highway Problem," *Railway Age,* n.s., 73 (Oct. 21, 1922): 760.

75. "Who Builds the Highways?" *Illinois Central Magazine* (Oct. 1922). Reprinted in *Railway Age,* n.s., 73 (Dec. 9, 1922): 1082.

76. Peter J. Hugill, "Good Roads and the Automobile in the United States, 1880–1929," *Geographical Review* 72 (July 1982): 349.

77. *Highway Motor Transportation,* xiii.

78. Lieb, *Transportation,* 57.

79. "Mileage: Truck vs. Train," derkeiler.com, http://newsgroups.derkeiler.com/Archive/Misc/misc.transport.trucking/2008-06/msg00035.html (accessed July 23, 2014).

80. James J. Flink, *The Automobile Age* (Cambridge, MA: MIT Press, 1988): 377–93.

81. Ibid., 404–9.

82. See Michael L. Berger, *The Automobile in American History and Culture: A Reference Guide* (Westport, CT: Greenwood Press, 2001), 17.

83. Ronald Primeau, *Romance of the Road: The Literature of the American Highway* (Bowling Green, OH: Bowling Green State University Popular Press, 1996): 1.

84. James P. Womack, Daniel T. Jones, and Daniel Roos, *The Machine That Changed the World* (New York: Rawson Associates, 1990): 11 and 38.

85. "This Third Largest Industry," *Automobile Topics* (Jan. 5, 1918): 948.

86. "Facts Worth Knowing," *Accessory And Garage Journal* 18 (Oct. 1928): 60.

87. "1928 Motor Vehicle Registrations" *Highway Transportation* 18 (May 1929): 20.

88. *Facts and Figures of the Automobile Industry,* 1931 ed., 1.

89. *Facts and Figures of the Automobile Industry,* 1928 ed. (New York: National Automobile Chamber of Commerce, 1928), 1.

90. *Facts and Figures of the Automobile Industry,* 1931 ed., 1.

91. *Fifteenth Census of the United States: 1930, Distribution,* vol. 1 (Washington, DC: United States Government Printing Office, 1933): 47.

92. "Growth of the Service Business," *Motor Age* 60 (Nov. 1941): 25.

93. "Annual Motor Vehicle Factory Sales, 1920–1960," *Automotive Industries* 104 (1961): 79.

94. James M. Rubenstein, *Making and Selling Cars: Innovation and Change in the U. S. Automotive Industry* (Baltimore: Johns Hopkins University Press, 2001), 331.

95. Womack, Jones, and Roos, *Machine That Changed the World,* 43.

96. Rubenstein, *Making and Selling Cars,* 343.

97. "U.S. Automotive Industry Employment Trends," Office of Aerospace and Automotive Industries, U.S. Department of Commerce, March 30, 2005, http://trade.gov/static/auto_reports_jobloss.pdf (accessed Aug. 5, 2014).

Chapter 3

1. "A New Inspection Carriage," *Railway Age* 33 (Feb. 28, 1902): 275.

2. "Rail Inspection Car," *Street Railway Journal* 24 (Dec. 3, 1904): 1018.

3. "A New Gasoline Motor Car," *Railway Age* 44 (Aug. 23, 1907): 271.

4. "Motor Cars for Maintenance of Way Forces," *Railway Age Gazette,* n.s., 55 (July 18, 1913): 102.

5. "Motor Cars for Section Forces," *Railway Age Gazette,* n.s., 50 (May 19, 1911): 1165.

6. Advertisement, *Railway Age Gazette,* n.s., 60 (June 30, 1916): 33b.

7. "Railroads Wise Will Motorize," *Railway Age,* n.s., 91 (Aug. 29, 1930): 314.

8. "Railroads Use Hundreds of Auto to Cut Costs," *Railway Age,* n.s., 93 (Aug. 6, 1932): 175.

9. Ibid., 179

10. Ibid., 178

11. "Railways Rapidly Adopting Emergency Auto Trucks," *Electric Railway Journal* 48 (July 22, 1916): 153.

12. "Automobiles Are Not Baggage," *Railway Age,* n.s., 30 (Nov. 30, 1900): 423.

13. "'Carbo' Steel Decks for Carrying Automobiles," *Railway Age,* n.s., 69 (July 16, 1920): 108.

14. "Michigan Central Has Another Good Year," *Railway Age,* n.s., 79 (July 11, 1925): 93.

15. *Facts and Figures of the Automobile Industry* (New York: National Automobile Chamber of Commerce, 1928), 14.

16. *Facts and Figures of the Automobile Industry* (New York: National Automobile Chamber of Commerce, 1934), 20.

17. *Motor Truck Facts* (Detroit: Automobile Manufacturers Association, 1940), 36.

18. "Automobile Baggage Car on Southern Pacific," *Railway Age,* n.s., 85 (Nov. 17, 1928): 971.

19. *Facts and Figures of the Automobile Industry* (1928), 14.

20. "Handling the Seminole Oil Business," *Railway Age,* n.s., 82 (May 7, 1927): 1405, 1407.

21. "Santa Fe Traffic Handles Rush Oil Field Traffic Without Delay," *Railway Age,* n.s., 82 (Mar. 26, 1927): 986.

22. "Railways Face New Competition," *Railway Age,* n.s., 89 (Nov. 15, 1930): 1022.

23. L. A. Rossman, *Railways and Highways: A Discussion of Transportation Relationships* (Grand Rapids, MN: Privately printed, 1935), 4.

24. R. W. Richardson, "The National Good Roads Association," *Good Roads Magazine* 33 (May 1903): 165.

25. "The Illinois Central Good Roads Movement," *Railroad Gazette*, n.s., 38 (Mar. 29, 1901): 211.

26. "The Good Roads Train on the Southern," *Railroad Gazette*, n.s., 38 (Nov. 8, 1901): 769.

27. "Plans for Automobile Railway," *Railway Age*, n.s., 52 (Mar. 29, 1907): 549.

28. "Suggests Replacing Obsolete Trackage With Roads," *Bus Transportation* 11 (Dec. 1932): 518.

29. "Making Taxicabs Operation Safer," *Electric Railroad Journal* 75 (Nov. 1931): 622.

30. "John D. Hertz," *Wikipedia*, http://en.wikipedia.org/wiki/John_D._Hertz (accessed Jan. 19, 2014).

31. "F. Parmelee & Co.," Coachbuilt.com, http://www.coachbuilt.com/bui/p/parmelee/parmelee.htm, pp. 4, 7 (accessed Jan. 19, 2014).

32. "Pennsylvania Opens New Passenger Station at Newark, N.J.," *Railway Age*, n.s., 98 (Mar. 30, 1935): 486.

33. Harris Saunders, *Top Up or Down? The Origin and Development of the Automobile and Truck Renting and Leasing Industry—56 Years, 1916–1972* (Birmingham, AL: Privately published, 1972), 143.

34. "Drummers Can Coordinate Rail and Automobile," *Railway Age*, n.s., 85 (July 14, 1928): 77.

35. "Railroad Will Have Auto Waiting," *Business Week* (Feb. 10, 1940): 18.

36. Advertisement for the Southern Railway in *Travel* 28 (Dec. 1916): 53.

37. "Has Ford Discovered an Ingenious Method of Rebating?" *Railway Age*, n.s., 77 (Aug. 1921): 408.

38. "The Work of the Ford Company's Traffic Department," *Railway Age*, n.s., 65 (Aug. 30, 1918): 402.

39. Ibid., 403.

40. Samuel Crowther, "Ford's Story of His Railway," *World's Work* 48 (June 1924): 161, 163.

41. "Mr. Ford's 'Railroad Miracle,'" *Railway Age*, n.s., 71 (Sept. 10, 1921): 473.

42. Daniel Willard, "Taxpayers, As Well As Railways, Are Victimized," *Railway Age*, n.s., 89 (Nov. 8, 1930): 971.

43. "A New Era for the Railroads," *Railway Age*, n.s., 90 (Feb. 21, 1931): 399.

44. "Inland Waters' Transport Costs Exceed Rail by 50 Per Cent," *Railway Age*, n.s., 89 (Nov. 22, 1930): 1077.

45. "A New Era for the Railroads," 400.

46. "The Subsidized Air Lines," *Railway Age*, n.s., 93 (Oct.–Dec. 1932): 822.

47. "Railways Handicapped in Meeting Highway Competition," *Railway Age*, n.s., 93 (Dec. 3, 1932): 811.

48. Ibid., 812.

49. Ibid., 814.

Chapter 4

1. "Grade Crossing Accidents," *Railway Age*, n.s., 68 (June 4, 1920): 1549.

2. "Safety Regulation at Railroad Grade Crossings," *Highway Transportation* 14 (July 1924): 11.

3. Alex Gordon, "Accidents at Grade Crossings and to Trespassers," *Railway Age Gazette* 60 (June 16, 1916): 1323; "Prevention of Accidents at Grade Crossings," *Railway Age Gazette* 60 (June 9, 1916): 1219.

4. "The Growing Menace of the Highway Crossing," *Railway Age,* n.s., 73 (Aug. 26, 1922): 1.

5. Leroy Scott, "The Railroads' Death-Toll," *World's Work* 9 (Jan. 1905): 5699.

6. Edward Bunnell Phelps, "America's Lead In Railroad Accidents," *World's Work* 15 (Nov. 1907): 9578–79.

7. *Railroad Gazette* 32 (1900): 424.

8. "Railroad Crossing Signboards in New York," *Railway Age,* n.s., 30 (July 20, 1900): 49.

9. "Damages for Track Elevation," *Railway Age,* n.s., 30 (Dec. 28, 1900): 507.

10. "Progress of Track Elevation in Chicago," *Railway Age,* n.s., 35 (Jan. 2, 1903): 23–28. The *Railway Age*'s later installments about the program of track elevation in Chicago, can be read in the following: "Track Elevation in Chicago," n.s., 10 (Jan. 1, 1904): 22–23; "Track Elevation of the Chicago & Western Indiana Through the City of Chicago," n.s., 37 (Mar. 24, 1905): 387, 389–92; and "Track Elevation on the Milwaukee Division, Chicago & Northwestern Railway," *Railway Age,* n.s., 44 (Aug. 16, 1907): 211, 213.

11. "Grade Crossings and Terminals in Washington," *Railroad Gazette* 32 (1900): 866.

12. "Grade Crossings," *Railroad Gazette* 32 (1900): 74.

13. "The Crossing Bell," *Railroad Gazette* 44 (July 6, 1900): 459.

14. "Highway Crossing Bells," *Railroad Gazette* 32 (July 20, 1900): 489.

15. J. S. Evans, "Highway Crossing Alarms," *Railroad Gazette* 32 (1900): 492.

16. "The Railroad and the Towns," *Railroad Gazette* 32 (1900): 568.

17. C.M.G., "Protection at Highway and Street Crossings of Railroads," *Railroad Gazette* 46 (Feb. 14, 1902): 107.

18. *Railway Age,* n.s., 36 (Sept. 4, 1903): 300.

19. *Railway Age,* n.s., 33 (Mar. 7, 1902): 277.

20. "Convention of Southern Railroad Commissioners," *Railway Age,* n.s., 34 (Oct. 24, 1902): 423.

21. "Track Elevation at Indianapolis," *Railway Age,* n.s., 35 (Feb. 20, 1903): 271.

22. *Railway Age,* n.s., 35 (Feb. 20, 1903): 275.

23. E. W. Vogel, "Preventing Highway Crossing Accidents," *Railway Age,* n.s., 35 (Mar. 20, 1903): 476.

24. "Wilson Railway Gate," *Railway Age Gazette* 55 (Mar. 18, 1910): 810; "An Automatic Crossing Gate," *Railroad Age Gazette* 54 (Aug. 20, 1909): 323–24.

25. "Motor Service on St. Joseph & Grand Island," *Railroad Age Gazette* 54 (Sept. 3, 1909): 421.

26. "Abolition of Grade Crossings on the Pennsylvania," *Railroad Gazette* 53 (Jan. 17, 1908): 101–2.

27. "General News Section," *Railroad Age Gazette* 54 (Jan. 8, 1909): 79.

28. John H. Brady, "Automobiles At Railroad Crossings," *Railroad Age Gazette* 45 (Sept. 11, 1908): 910.

29. "Cost of Grade Crossings on Long Island," *Railway Age Gazette* 51 (Sept. 8, 1911): 485.

30. "Why the Railroads Kill," *World's Work* 15 (Mar. 1908): 9960.

31. William J. Wilgus, "Proposed New Freight Subway; New York City And Port," *Railroad Age Gazette* 45 (Oct. 16, 1908): 1150, 1157.

32. "Annual Report of New York State, Public Service Commission, Second District," *Railway Age Gazette* 50 (Jan. 20, 1911): 120.

33. "Grade Crossings for New York City," *Railway Age Gazette* 53 (Nov. 1, 1913): 850.

34. "National Association of Railway Commissioners," *Railway Age Gazette* 61 (Nov. 17, 1916): 898–99.

35. "The Truth About Railway Accidents," *Railway Age Gazette* 51 (Dec. 8, 1911): 1166–70.

36. "Intelligent Agitation of the Grade Crossing Problem," *Railway Age Gazette* 52 (Jan. 19, 1912): 81–82.

37. "Maintenance of Way Section," *Railway Age Gazette* 52 Mar. 15, 1912): 485.

38. "Automobile Innocent," *Railway Age Gazette* 53 (Nov. 15, 1912): 962.

39. "No Railroad Negligence on Grade Crossings," *Railway Age Gazette* 60 (June 18, 1913): 1384. See also "Automobile Accidents at Railroad Crossings Comparatively Few," *Railway Age Gazette* 55 (July 11, 1913): 64; "The Cause of Automobile Accidents at Highway Crossings," *Railway Age Gazette* 55 (Nov. 21, 1913): 991–92.

40. "Drastic Grade Crossing Law in New Jersey," *Railway Age Gazette* 54 (Mar. 14, 1913): 515.

41. "Grade Crossing Elimination," *Railway Age Gazette* 59 (Oct. 8, 1915): 634–35.

42. "To Promote Safety at Highway Crossings," *Railway Age Gazette* 59 (Dec. 17, 1915): 1119.

43. "Illinois Central Crossing Accident Campaign," *Railway Age Gazette* 61 (Dec. 8, 1916): 1061.

44. "Jury Blames Auto Driver for Accident," *Railway Age Gazette* 63 (July 13, 1917): 56.

45. "A Highway Crossing Alarm with Four Indications," *Railway Age Gazette* 58 (Mar. 19, 1915): 677–78; "Flash-Light Crossing Signal," *Railway Age Gazette* 62 (May 18, 1917): 1061–62; "Semaphores for Highway Crossings," *Railway Age Gazette* 62: (May 18, 1917): 1040.

46. "Grade Crossings and Automobiles," *Railway Age Gazette* 55 (Aug. 29, 1913): 360–62.

47. "Proposed Regulation of Automobiles," *Railway Age Gazette* 56 (Mar. 13, 1914): 497.

48. "Legislation to Get Around the Courts," *Railway Age Gazette* 56 (May 1, 1914): 73–74.

49. "Grade Crossing Law in New Jersey," *Railway Age Gazette* 59 (May 1, 1914): 997.

50. "Not Bound to Warn Travelers," *Railway World* 58 (Aug. 1914): 670.

51. "Safety First on the Long Island," *Railway Age Gazette* 58 (June 11, 1915): 140; "The Automobile and the Grade Crossing Problem," *Railway World* 59 (July 1915): 546; "Warning Posters for Automobilists on Long Island," *Railway Age Gazette* 60 (June 2, 1916): 1193.

52. "Long Island Grade Crossing Campaign," *Railway Age Gazette* 60 (June 16, 1916): 1342; "The Long Island's Life Saving Bulletins," *Railway Age*, n.s., 61 (Sept. 1, 1916): 382.

53. "Good Service at Highway Crossings," *Railway Age Gazette* 60 (Jan. 14, 1916): 45–46.

54. "Joint Grade Crossing Report," *Railway Age Gazette* 61 (July 28, 1915): 157.

55. "Summit-Hallstead Cut-Off of D.L. & W.," *Railway Age Gazette* 55 (Dec. 5, 1913): 1069–74.

56. "Lackawanna Improvements in Orange, N.J.," *Railway Age*, n.s., 65 (Nov. 15, 1918): 845.

57. E. L. Wonson, "Elimination of the Tower Grove Crossings, St. Louis," *Railway Age Gazette* 59 (Oct. 29, 1915): 799–802.

58. "When the Shoe Pinches the Other Foot," *Railway Age Gazette* 61 (Oct. 6, 1916): 1.

59. "Highway Crossing Signals," *Railway Age Gazette* 61 (Nov. 24, 1916): 927.

60. "Automobile Accidents at Grade Crossings," *Railway Age*, n.s., 67 (Aug. 15, 1919): 327.

61. "Better Protection for Highway Crossings," *Railway Age*, n.s., 65 (Aug. 16, 1918): 286.

62. "Uniform Highway Crossing Signs in Connecticut," *Railway Age*, n.s., 65: 26 (Dec. 17, 1918): 1164.

63. "Safety At Highway Grade Crossings," *Railway Age Gazette* 63 (Nov. 2, 1917): 804.

64. "Concrete Highway Crossings," *Railway Age,* n.s., 64 (Jan. 25, 1918): 215–16.

65. "'Doing Her Bit,'" *Railway Age Gazette* 62 (May 18, 1917): 1062; "Doing Her Bit," *Train Dispatchers Bulletin* 22 (July 1917): 19.

66. "Elimination of Grade Crossings in Chicago," *Railway Age,* n.s., 67 (Dec. 9, 1919): 991.

67. "A Means of Protecting Grade Crossings," *Railway Age,* n.s., 69 (Dec. 31, 1920): 1162–63.

68. "Safety Regulations at Railroad Crossings," *Railway Age,* n.s., 76 (Feb. 23, 1924): 462.

69. "Automobile Accidents at Grade Crossings," *Railway Age,* n.s., 70 (Feb. 7, 1925): 350–51. See also A. H. Rudd, "Protection of Highway Traffic at Crossings," *Railway Age,* n.s., 76 (Mar. 8, 1924): 557–58; and Robert H. Ford, "Can the Grade Crossing Problem Be Solved?" *Railway Age,* n.s., 76 (May 31, 1924): 1332–33.

70. E. Irvine Rudd, "Accidents and Protective Devices at Grade Crossings," *Railway Age,* n.s., 73 (Sept. 22, 1928): 561–62.

71. "Automobile Accidents at Grade Crossings," *Railway Age,* n.s., 70 (Feb. 7, 1925): 351.

72. "Report of Grade Crossing Committee," *Railway Age,* n.s., 88 (May 31, 1930): 1321.

73. "How the Erie Has Reduced Highway Crossing Accidents," *Railway Age,* n.s., 92 (Aug. 18, 1934): 215.

74. "Another Warning to Careless Motorists," *Electric Railway Journal* 64 (Dec. 13, 1921): 1012.

75. "The Careful Crossing Campaign in Chicago," *Railway Age,* n.s., 73 (Oct. 7, 1922): 671. See also "Arouse the Public," *Railway Age,* n.s., 76 (June 3, 1921): 1252; and "Highway Crossing Accidents," *Railway Age,* n.s., 80 (June 5, 1926): 1468.

76. "A Safety Message by Radio," *Railway Age,* n.s., 72 (June 3, 1922): 1302.

77. "Automobile Disasters Unchecked—Need of a More Extensive Campaign," *Railway Age,* n.s., 73 (Oct. 7, 1922): 644.

78. "To Investigate Automobile Accidents," *Railway Age,* n.s., 73 (July 15, 1922): 79.

79. "The N.Y.C's 'Canned' Safety Speech," *Railway Age,* n.s., 73 (July 15, 1922), 121.

80. "Results of Careful Crossing Campaign," *Railway Age,* n.s., 73 (Sept. 8, 1922): 451.

81. Floyd W. Parsons, "Saving Men and Money," *World's Work* 47 (Nov. 1923): 33.

82. "'Be Careful at Crossings,'" *Railway Age,* n.s., 79 (July 4, 1925): 48.

83. "Supreme Court on the Paterson Crossing Case," *Railway Age,* n.s., 70: 4 (Jan. 28, 1921): 268.

84. "Northern Pacific Illuminated Highway Signs," *Railway Age,* n.s., 71 (Nov. 12, 1921): 946.

85. "Resist the Opening of New Grade Crossings," *Railway Age,* n.s., 77 (Dec. 20, 1924): 1105–6.

86. C. F. Loweth, "The Railroad and the City Plan," *Railway Age,* n.s., 80 (June 5, 1926): 1478.

87. "Development of Concrete in Railway Construction: A Study of the Design of the Minor Structures on the Delaware, Lackawanna & Western" *Railway Age,* n.s., 73 (Oct. 28, 1922): 791–97.

88. "Report on Signs, Fences and Crossings," *Railway Age,* n.s., 74 (Mar. 14, 1923): 635–36.

89. "Big Four Cutoff Involves Monumental Bridge," *Railway Age,* n.s., 77 (June 19, 1924): 99 and 102–3.

90. "A Program for Fewer Deaths at Railroad Crossings," *American City Magazine* 31 (Sept. 1924): 186.

91. "Types of Pavement for Railway Crossings," *American City Magazine* 32 (Jan. 1925): 11.

92. "Automatic Highway Crossing Signals in Place of Gates," *Railway Age,* n.s., 73 (Nov. 4, 1922): 835–36.

93. "Possible Methods of Protection," *Railway Age,* n.s., 73 (Oct. 7, 1922): 642–44.

94. A. H. Rudd, "Protection of Highway Traffic at Crossings," *Railway Age,* n.s., 76 (Mar 8, 1924): 558.

95. T. R. Ratcliff, "Crossing Gates Replaced by Flashing-Light Signals," *Railway Age,* n.s., 73 (Sept. 1, 1928): 426; "Railroad Grade Crossings Protected by Three-Color Electric Traffic Signals," *American City* 38 (June 1928): 151–52.

96. "Automatic Crossing Gates on the I. C.," *Railway Age,* n.s., 98 (Feb. 16, 1935): 271–72.

97. "'N.C. Stop Law' Is Now in Effect," *Railway Age,* n.s., 68 (July 11, 1923): 55.

98. "State Laws Regulating Vehicles at Crossings," *Railway Age,* 75 (Nov. 3, 1923): 825–26.

99. "The Protection of Railroad Grade Crossings," *Railway Age,* n.s., 77 (Sept. 20, 1924): 514.

100. "Recommendations for Prevention of Crossing Accidents," *Railway Age,* n.s., 77 (Nov. 29, 1924): 995–96.

101. "The Grade Crossing Problem," *Railway Age,* n.s., 77 (July 5, 1924): 28. See also "Whose Problem Is Grade Separation?" *Railway Age,* n.s., 76 (Apr. 12, 1924): 916–17.

102. "Crossing Signals Erected by Bus Line," *Bus Transportation* 7 (May 1928): 289.

103. Howard A. Shieber, "The Deadly Grade Crossing," *American Review of Reviews* 74 (Nov. 1926): 522. See also Robert H. Ford, "Who Should Pay for Grade Crossing Separation?", *Railway Age,* n.s., 74 (Apr. 27, 1929): 961–62.

104. Robert H. Ford, "Lack of Unified Control Complicates Highway Crossing Problems," *Railway Age,* n.s., 82 (May 7, 1927): 1385.

105. "Real Progress Being Made in Federal Grade Crossing Program," *Railway Age,* n.s., 101 (Sept. 12, 1936): 383.

106. Charles W. Galloway, "Grade Separation—Is It the Solution For Crossing Accidents?" *Railway Age,* n.s., 75 (June 21, 1930): 1474.

107. "Oil Truck Wrecks at Crossings are Numerous," *Railway Age,* n.s., 95 (Aug. 26, 1933): 312–13.

108. "Two Views on Grade Crossing Problem," *Railway Age,* n.s., 98 (May 11, 1935): 739–42.

109. Thomas H. MacDonald, "The Grade Crossing problem," *Railway Age,* n.s., 102 (Mar. 27, 1937): 545–47.

110. "Grade Crossing Accidents," *Railway Age,* n.s., 88 (May 10, 1930): 1137.

111. "Modern Highway Crossing Protection Reduces Operating Costs," *Railway Age,* n.s., 91 (Nov. 14, 1931): 737–38.

112. "A New Approach to the Grade Separation Problem," *Railway Age,* n.s., 93 (Oct. 8, 1932): 495.

113. "Modern Highway Crossing . . ." *Railway Age,* n.s., 93 (Oct. 8, 1932): 494–95.

114. Harry D. Blake, "Combatting Unemployment with a Grade Separation Program," *Railway Age,* n.s., 91 (Dec. 12, 1931): 896–99.

115. "Extensive Grade Separation Project Has Many Points of Interest," *Railway Age,* n.s., 97 (Sept. 15, 1934): 305.

116. "Grade Crossings to Be Eliminated in City of Syracuse in New York Central Project," *Railway Age,* n.s., 96 (Apr. 21, 1934): 571–74.

117. "Grade Crossing Projects Reported to P.W.A.," *Railway Age,* n.s., 98 (Mar. 2, 1933): 338.

118. "$200,000,000 for Grade Separation," *Railway Age,* n.s., 98 (June 22, 1935): 969.

119. "Federal Funds for Highway Crossing Protection," *Railway Age*, n.s., 98 (Mar. 9, 1935): 356. See also John H. Dunn, "Crossing Protection Installations Increased During 1935," *Railway Age*, n.s., 96 (Jan. 4, 1936): 83–86; and "Chicago & North Western Installs Automatic Gates," *Railway Age*, n.s., 101 (Oct. 3, 1936): 470–72.

120. Robert H. Ford, "The Highway Grade Crossing—A National Problem," *Railway Age*, n.s., 98 (Apr. 27, 1935): 643–45.

121. "New York Railroad Club Discusses the Grade Crossing Problem," *Railway Age*, n.s., 100 (May 23, 1936): 829–32.

122. "Federal Grade Crossing Program Is in Full Swing," *Railway Age*, n.s., 102 (Mar. 13, 1937): 425, 429.

123. "Federal Grade Crossing Program Moves Steadily Forward," *Railway Age*, n.s., 108 (Mar. 11, 1939): 407–9.

124. J. C. Furnas, "Gray Hairs for Casey Jones," *Saturday Evening Post*, Apr. 10, 1937, 81.

125. "Auto-Stop Installed on C. & N.W.," *Railway Age*, n.s., 99 (Aug. 24, 1935): 239–40.

126. "Real Progress Being Made in Federal Grade Crossing Program," *Railway Age*, n.s., 101 (Sept.12, 1936): 383.

127. "Illumination for Grade Crossings," *Railway Age*, n.s., 104 (Apr. 24, 1937): 719–20; "Illumination of Grade Crossing," *Railway Age*, n.s., 104 (Apr. 24, 1937): 703. Despite their similar titles and the same publication date, these are two separate articles.

128. *Railway Age*, n.s., 108 (Mar. 9, 1940): 434.

129. Millard Matovina, "Extensive Crossing Protection Ordered by the City of Gary, Ind." *American City* 63 (Aug. 1948): 131, 133. See also "Signals for Trains at Highway Crossings," *Railway Age*, n.s., 95 (Dec. 23, 1950): 25–26.

130. John Stilgoe, *Metropolitan Corridor: Railroads and the American Scene* (New Haven, CT: Yale University Press, 1983), 188.

131. *Shanklin v. Norfolk Southern Railway Company*, U.S. 6th Circuit. Available online at Findlaw.caselaw.com (accessed Aug. 4, 2014).

132. Robert L. Pottroff, "Crossing Accidents: Who's Getting Railroaded Now?" Pottroff Law Office, P.A., website, http://www.pottroff.com/publications/accidents-railroaded-now (accessed Jan. 19, 2014).

133. "Tracks, Trains and Automobiles: Safety at Railroad Grade Crossings," *Tech Transfer*, University of California, Berkeley, http://techtransfer.berkeley.edu/newsletter/04-4/tracks.php (accessed Jan. 19, 2014).

134. Shannon Mok and Ian Savage, "Why Has Safety Improved at Rail-Highway Grade Crossings?" *Proceedings of the Transportation Research Forum* (2004): 3. Available online at http://www .trforum.org/forum/downloads/2004_ImproveSafety_paper.pdf (accessed Jan. 26, 2016).

135. Ibid, 5.

136. Ibid., 3–4.

Chapter 5

1. For a brief history of the electric railway, see "Electric Railway Work in America Prior to 1888," *Street Railway Journal* 24 (Oct. 8, 1904): 559–60. For a brief overview of the horse car era by

one who participated in it, see C. Densmore Wyman, "The Evolution Of The Industry In Horse Car Days," *Street Railway Journal* 24 (Oct. 8, 1904): 525–26.

2. "Forty Years of 'Electric Railway Journal,'" *Electric Railway Journal* 64 (Dec. 6, 1924): 947.

3. Harry L. Brown, "A Background of Forty Years for Future Building," *Electric Railway Journal* 64 (Dec. 6, 1924): 945.

4. "Forty Years of 'Electric Railway Journal,'" *Electric Railway Journal* 64 (Dec. 6, 1924): 950.

5. "Street Railways During the Last Decade," *Street Railway Journal* 24 (Oct. 8, 1904): 599–600.

6. Robert McCulloch, "St. Louis Street Railways in 1884 and 1904," *Street Railway Journal* 24 (Oct. 8, 1904): 580–81.

7. "Notes on Los Angeles Railway Company's System," *Street Railway Journal* 23 (Apr. 9, 1904): 552–54.

8. "Automobiles and Street Cars," *Street Railway Journal* 22 (June 4, 1904): 829.

9. "Trolley Or Automobile?" *Literary Digest* 30 (May 6, 1905): 660.

10. C. E. Smith, "City Transportation and Street Congestion," *Electric Railway Journal* 64 (Dec. 6, 1924): 960.

11. "Relieving Traffic Congestion in Large Cities," *Electric Railway Journal* 44 (Sept. 26, 1914): 551.

12. "Automobiles In City Traffic," *Electric Railway Journal* 44 (Dec. 12, 1914): 1.

13. "Safety a Real Economy in Columbus," *Electric Railway Journal* 48 (Apr. 22, 1916): 772, 774.

14. H. G. Windsor, "Motor Vehicle Accidents and Traffic Regulation," *Electric Railway Journal* 48 (Oct. 21, 1916): 874–77.

15. "Electric Railway Section of National Safety Council," *Electric Railway Journal* 48 (Mar. 18, 1916): 564.

16. "Chicago's Congestion Problem," *Electric Railway Journal* 48 (Aug. 12, 1916): 273. The other two articles in the Chicago trilogy include "Study of Chicago's Congested Traffic," *Electric Railway Journal* 48 (July 29, 1916): 183–88; and "Chicago's Congested Streets," *Electric Railway Journal* 48 (Aug. 5, 1916): 218–21.

17. "Speeding up Traffic in Dallas," *Electric Railway Journal* 52 (Dec. 7, 1918): 1001.

18. Ibid., 1000.

19. "Chicago Loop Rerouting Would Increase Capacity for Cars 30%," *Electric Railway Journal* 63 (May 3, 1924): 683–87.

20. "Relief from Congestion Suggested in Chicago," *Electric Railway Journal* 63 (June 7, 1924): 907.

21. "The Space-Grabbing Automobile," *Electric Railway Journal* 52 (July 27, 1918): 147.

22. "Parked Autos Steal Streets," *Electric Railway Journal* 51 (June 1, 1918): 1040.

23. J. M. Quigley, "Traffic Regulations and Safety Work," *Electric Railway Journal* 58 (Oct. 8, 1921): 607.

24. "Safety Zone Plan Extended in Los Angeles," *Electric Railway Journal* 57 (June 18, 1921): 1149.

25. "Report on Rapid Transit System for Detroit," *Electric Railway Journal* 51 (Mar. 2, 1918): 420.

26. C. K. McGunnegle and M. T. Montgomery, "What Automobile Traffic Means to Pittsburgh," *Electric Railway Journal* 58 (July 2, 1921): 21.

27. "Observer: Des Moines Rides Buses and Walks," *Electric Railway Journal* 58 (Aug. 20, 1921): 287–88.

28. Ibid., 284.

29. "Lessons Learned by Des Moines Citizens," *Electric Railway Journal* 58 (Sept. 10, 1921): 1017.

30. John H. Hanna, "Evolution of Community Transportation," *Electric Railway Journal* 75 (Sept. 15, 1931): 501, 502.

31. Morris Buck, "Financial Requirements Met Without Difficulty," *Electric Railway Journal* 75 (Jan. 1931): 22.

32. Edwin Gruhl, "The Traction Industry Today and Four Years Ago," *Electric Railway Journal* 58 (Oct. 8, 1921): 595, 599.

33. Ibid., 595, 597.

34. Ibid., 599.

35. L. F. Eppich, "Electric Railways from Realtor's Standpoint," *Electric Railway Journal* 64 (Oct. 11, 1924): 628–29.

36. Clinton E. Morgan, "Making Transportation Serve—and Sell," *Electric Railway Journal* 58 (Sept. 24, 1921): 499.

37. Frank H. Warren, "How Can Salesmanship Be Applied in the Street Railway Business?" *Electric Railway Journal* 58 (Dec. 3, 1921): 987.

38. "Anticipate the Automobile Age in Your Building Plans," *Electric Railway Age* 57 (Jan. 8, 1921), 69.

39. C. E. Smith, "City Transportation and Street Congestion," *Electric Railway Journal* 64 (Dec. 6, 1924): 961–63.

40. "Speeding up Traffic in Detroit," *Electric Railway Journal* 63 (June 14, 1924): 927, 929.

41. "D.U.R. to Operate Co-ordinated Rail-Bus System," *Electric Railway Journal* 64 (Nov. 22, 1924): 879.

42. "Track Removal Will Not Relieve Congestion," *Electric Railway Journal* 63 (May 31, 1924): 858.

43. Howard F. Fritch, "What Railway Operators Can Do to Improve Traffic Conditions," *Electric Railway Journal* 63 (May 31, 1924): 845–46.

44. J. W. McCloy, "Rolling Stock Purchases Continue at Moderate Rate," *Electric Railway Journal* 75 (Jan. 1931): 40; John Anderson Miller, *Fares, Please! From Horse-Cars To Streamliners* (New York: D. Appleton-Century, 1941): 110. See also "More Open Cars Converted for One-Man Operation," *Electric Railway Journal* 57 (June 25, 1921): 1167; this article discusses the Eastern Massachusetts Street Railways' earlier reconstruction of open-sided cars into close-sided ones, along with other improvements.

45. G. A. Richardson, "Looking Ahead in Urban Transportation," *Electric Railway Journal* 75 (Sept. 15, 1931): 504.

46. "Automotive Principles Applied in Design of New Blackhall Car," *Electric Railway Journal* 75 (Feb. 1931): 75.

47. Ibid., 74.

48. J. W. McCloy, "Rolling Stock Purchases Continue at Moderate Rate," *Electric Railway Journal* 75 (Jan. 1931): 43.

49. Ibid., 40.

50. John A. Miller, "Car Design Reflects Steadily Rising Standards of Service," *Electric Railway Journal* 75 (Sept. 15, 1931): 518.

51. "Detroit Survey Develops Basic Traffic Data: Part One," *Electric Railway Journal* 71 (Apr. 1928): 571.

52. "Detroit Survey Develops Basic Traffic Date: Part Two," *Electric Railway Journal* 71 (Apr. 14, 1928): 617–21.

53. "Redistribution of Street Space Proposed at Detroit," *Electric Railway Journal* 75 (July 1931): 362.

54. "Surface Rapid Transit Line Operated in Detroit," *Electric Railway Journal* 71 (Jan. 7, 1928):4.

55. E. J. McIlraith, "Chicago Prohibits Parking in Its Central Area," *Electric Railway Journal* 71 (Feb. 4, 1928): 191.

56. "Rebuilding Track Under Heavy Traffic," *Electric Railway Journal* 75 (Dec. 1931): 685–88.

57. "Safety a Science at Louisville," *Electric Railway Journal* 71 (Jan. 28, 1928): 149–51.

58. "Diagonal Versus Parallel Parking," *Electric Railway Journal* 72 (Dec. 22, 1928): 1079.

59. "Higher Maintenance Standards and Lower Costs Are Objectives at Atlanta," *Electric Railway Journal* 75 (Nov. 1931): 627.

60. "Public Sentiment Favors Loading Platforms in Cincinnati," *Electric Railway Journal* 75 (Dec. 1931): 644.

61. "Lax Enforcement of Traffic Rules: A Prevalent Cause of Congestion," *Electric Railway Age* 75 (Aug. 1931): 400–404.

62. *Census of Electrical Industries, Street Railways and Trolley-Bus And Motorbus Operations* (Washington, D. C.: Government Printing Office, 1939): 3–4.

63. G. A, Richardson, "Looking Ahead in Urban Transportation," *Electric Railway Journal* 75 (Sept. 15, 1931): 504, 506.

64. H. Roger Grant, *Getting Around: Exploring Transportation History* (Malabar, FL: Krieger Publishing Company, 2003): 139.

65. William D. Middleton, *The Time of the Trolley* (Milwaukee, WI: Kalmbach Publishing, 1967): 397.

66. "Jitney Situation on the Pacific Coast," *Electric Railway Journal* 42 (Mar. 11, 1916): 497–98.

67. "Effect of Jitney Operation in Seattle," *Electric Railway Journal* 47 (Feb. 5, 1916): 291.

68. "Jitneys as Steam Railway Competitors," *Electric Railway Journal* 43 (Oct. 14, 1916): 857.

69. Edwin Gruhl, "The Traction Industry Today and Four Years Ago," *Electric Railway Journal* 58 (Oct. 8, 1921): 597.

70. Delos F. Wilcox, *Analysis of the Street Railway Problem* (New York: Delos F. Wilcox, 1921), 112.

71. "Jitneys Fall Off in Kansas City," *Electric Railway Journal* 47 (Mar. 4, 1916): 445.

72. "Jitneys Off Railway Streets," *Electric Railway Journal* 57 (Apr. 16, 1921): 700.

73. "Non-Essential Jitneys Must Go," *Electric Railway Journal* 52 (Oct. 26, 1918): 745.

74. A. M. Welch, "The 'Jitney' Bus," *Highway Transportation* 11 (Sept. 1921): 22.

75. "Jitneys Under State Control in Connecticut," *Electric Railway Age* 57 (Apr. 1921): 792.

76. "Tell This to the Motorists Who Pick Up Railway Passengers," *Electric Railway Journal* 63 (June 7, 1924): 1.

77. "Jitneys Return at Davenport," *Electric Railway Journal* 57 (Feb. 5, 1921): 286.

78. "An Autoist Applies the Golden Rule," *Electric Railway Journal* 48 (Oct. 14, 1916): 856–57.

79. "Surface Rapid Transit Line Operated in Detroit," *Electric Railway Journal* 71 (Jan. 7, 1928): 7.

80. "Bus Service to Replace Detroit's Jitneys," *Electric Railway Journal* 72 (Dec. 1, 1928): 977.

81. "The Service Car Challenge," *Bus Transportation* 11 (July 1932): 312.

82. Th. M. Vanderstempel, "Miami Transportation Triangle," *Bus Transportation* 17 (May 1938): 214–15.

83. Gardner W. Pearson, "Are the Trolleys the Only Practical System of Transportation?" *Electric Railway Journal* 58 (Nov. 12, 1921): 866.

84. Miller, *Fares, Please,* 154–55.

85. R. E. Fulton, "The Motor Bus vs. The Street Car," *Highway Transportation* 51 (Dec. 1919): 17.

86. Ibid., 18. See also G. T. Seely, "How the Motor Coach Can Be Utilized in the Transportation System," *Electric Railway Journal* 63 (Apr. 5, 1924): 545.

87. C. W. Stocks, "The Bus Transportation Field" *Electric Railway Journal* 58 (Sept. 24, 1921): 519.

88. Pearson, "Are the Trolleys the Only Practical System of Transportation?" 867.

89. Ibid., 866.

90. William P. Kennedy, "The Motor Bus—Keynote of Expansion for Street Railways," *Electric Railway Journal* 63 (Jan. 26, 1924): 145.

91. J. A. Emery, "Co-ordination of Bus and Trolley," *Electric Railway Journal* 63 (Feb. 9, 1924): 211–13.

92. R. R. Blackwell, "Another Milestone," *Highway Transportation* 11 (Sept. 1921): 16.

93. Emery, "Co-ordination of Bus and Trolley," 211; Luke Grant, "How the Motor Coach Can Be Utilized in the Transportation System," *Electric Railway Journal* 63 (Apr. 5, 1924): 546.

94. Carl W. Stocks, "The Modern Motor Bus," *Electric Railway Journal* 75 (Sept. 15, 1931): 547.

95. H. B. Titcomb, "The Bus and Trolley," *Electric Railway Journal* 57 (Jan. 8, 1921): 81.

96. "Akron's Motor Bus Route," *Electric Railway Journal* 57 (June 25, 1921): 1165.

97. "Akron Company Had No Other Course but to Cease Operation," *Electric Railway Journal* 63 (Feb. 9, 1924):201–2; "Akron 'Rubber Urge' Passes," *Electric Railway Journal* 63 (Mar. 1, 1924): 330–32.

98. "Detroit Answers Motorbus Proposal," *Electric Railway Journal* 63 (Mar. 29, 1924): 501–2.

99. "Detroit Extends Surface Rapid Transit System," *Electric Railway Journal* 72 (Dec. 15, 1928): 1036–38.

100. "De Luxe Bus Lines Now Serve Detroit," *Electric Railway Journal* 72 (Dec. 22, 1928): 1075–79.

101. "Value of the Bus," *Bus Transportation* 7 (Feb. 1928): 68–69.

102. "Expansion of Bus Operation," *Electric Railway Journal* 71 (Jan 14, 1928): 63 and 68–69.

103. "90-Day Trial Clinched Change to Buses," *Bus Transportation* 10 (Dec. 1931): 619–21.

104. "The Taxi-Coach: A New Development in the Bus Field," *Electric Railway Journal* 75 (May 1931): 263.

105. "Eau Claire: Another Recruit in the All-Bus Army," *Bus Transportation* 12 (May 1933): 193–94; "New Deal with a Fresh Deck: Trenton Goes All Bus," *Bus Transportation* 13 (Dec. 1934): 456–57; "Buses to Replace Trolleys at Green Bay, Wisconsin," *Bus Transportation* 16 (Oct. 1937): 508; "Charleston, S.C. Goes all Bus," *Bus Transportation* 17 (Mar. 1938): 151–52; "Harrisburg, Pa., Joins All-Bus Rank," *Bus Transportation* 18 (Aug. 1939): 428.

106. For example, see "Honolulu Gets a New Deal," *Bus Transportation* 13 (Mar. 1934): 88–89.

107. "How Cities Are Served," *Transit Journal* 85 (Jan. 1941): 8–9.

108. "Cities on the Toboggan," *Transit Journal* 85 (Feb. 1941): 47.

109. "Comments about 'Cities on the Toboggan,'" *Transit Journal* 85 (Apr. 1941): 143–45.

110. "Transit's Emergency Job," *Transit Journal* 85 (Apr. 1941): 122–25; Robert G. Silar, "Detroit Runs 'Zone Buses' For War Workers," *Transit Journal* 85 (May 1941): 160–61.

111. For a history of the trackless trolley before 1900, see Miller, *Fares, Please,* 165–72.

112. "The Trackless Trolley," *Street Railway Journal* 24 (Nov. 12, 1904): 866–67.

113. "Trackless Trolley vs. Motor Omnibus," *Electric Railway Journal* 47 (Mar. 4, 1916): 444.

114. C. W. Stocks, "The Bus Transportation Field," *Electric Railway Journal* 58 (Sept. 24, 1921): 517.

115. "Another Advantage of Electrification Emphasized," *Electric Railway Journal* 57 (June 24, 1921): 1161.

116. "Pioneer Trackless Trolley Installation," *Electric Railway Journal* 57 (June 25, 1921): 1158.

117. "Another Trolley Bus Tested," *Electric Railway Journal* 58 (Sept. 24, 1921): 522. See also "Trackless Trolley Tested at Schenectady," *Electric Railway Journal* 57 (June 18, 1921): 1132.

118. "Trolley Bus Operation Commences on Staten Island," *Electric Railway Journal* 58(Oct. 15, 1921): 689.

119. "Trackless Trolleys for New York," *Electric Railway Journal* 57 (May 28, 1921): 1002.

120. "Trackless Trolleys Give Improved Service to Cohoes, N.Y.," *Electric Railway Journal* 64 (Dec. 13, 1924): 985.

121. "Trackless Trolley New Development in Passenger Transportation," *Highway Transportation* 11 (Oct. 1921): 11. See also "Details of Packard Trolley Bus," *Electric Railway Journal* 58 (Sept. 3, 1921): 359–60.

122. "Eyes Are Again on Richmond: This Time the Trackless Electric Car," *Electric Railway Journal* 57 (June 25, 1921): 1156.

123. "Seeing the Trolley Bus in Proper Perspective," *Electric Railway Journal* 75 (Feb. 1931): 65.

124. "The Trackless Trolley: Has It a Place?" *Electric Railway Journal* 72 (Sept. 8, 1928): 429–30.

125. E. A. West, "Salt Lake City Chose the Bus," *Bus Transportation* 14 (Mar. 1935): 106.

126. "Letter to the Editor," *Bus Transportation* 14 (July 1935): 277.

127. "Trolley Bus Experiences Unprecedented Growth During Past Year," *Electric Railway Journal* 75 (Jan. 1931): 45.

128. Ibid., 48.

129. "Trolley Bus Expansion Continues," *Transit Journal* 85 (Jan. 1941): 11.

130. Kent S. Putnam, "More Trolley Buses in More Cities," *Transit Journal* 85 (Dec. 1941): 454, 456.

Chapter 6

1. W. H. Perry, "Street Railways vs. Good Road," *Good Roads Magazine* 32 (Aug. 1901): 8, 9.

2. George W. Hilton and John F. Due, *The Electric Interurban Railways in America* (Stanford, CA: Stanford University Press, 1960), 186.

3. Ibid., 186–87.

4. Ibid., 196–97.

5. Ibid., 12, 41.

6. Ibid., 42.

7. "Ohio Interurbans Develop Through Traffic," *Street Railway Journal* 24 (Aug. 20, 1904): 263.

8. "The Dayton, Covington & Piqua Company," *Street Railway Journal* 24 (Sept. 17, 1904): 390.

9. F. W. Doolittle, "The Present and Future Development of Interurban Railways," *Electric Railway Journal* 48 (Sept. 2, 1916): 393.

10. Henry Marlette, *Electric Railroads of Indiana* (Indianapolis: Hoosier Heritage Press, 1980), 37.

11. Hilton and Due, *Electric Interurban Railways*, 50.

12. Theodore Stebbins, "Some Notes on the Evolution of Electric Transportation," *Proceedings of the American Street and Interurban Railway Association* (1906): 261.

13. Ibid., 94 .

14. Hilton and Due, *Electric Interurban Railways*, 95–96.

15. Doolittle, "Present and Future Development," 393.

16. Hilton and Due, *Electric Interurban Railways*, 55.

17. "Selling Energy Along Interurban Railways," *Electric Railway Journal* 48 (Oct. 28, 1916): 922.

18. Ibid. 921.

19. "The Indiana Northern Traction Company," *Street Railway Journal* 24 (Sept. 10, 1904): 365.

20. "Sleeping Car Service Begun by Interstate Public Service Company," *Electric Railway Journal* 64 (Nov. 8, 1924): 802.

21. "Indianapolis Traction Terminal Building," *Street Railway Journal* 24 (Sept. 10, 1904): 869–90.

22. Bernard Meiklejohn, "Electricity Transforming Traffic," *World's Work* 10 (May 1905): 6183.

23. "Operating Features of Norumbega Park," *Electric Railway Journal* 43 (Feb. 21, 1914): 411.

24. B. M. Colt, "Old and New Features at Sacandaga Park," *Electric Railway Journal* 43 (Apr. 25, 1914): 927.

25. "Many Factors Contributed to Receivership," *Electric Railway Journal* 57 (Feb. 26, 1921): 421.

26. Hilton and Due, *Electric Interurban Railways*, 233.

27. Ibid., 234.

28. Ibid., 248.

29. Thomas Conway, Jr., "Outlook for Interurbans Stated," *Electric Railway Journal* 63 (Feb. 2, 1924): 182.

30. "Buses that Compete with Rails Barred," *Bus Transportation* 15 (Apr. 15, 1936): 183.

31. "The 'Electric Railway Journal' and Bus Transportation," *Electric Railway Journal* 58 (Oct. 29, 1921): 764.

32. "Electric Railways Operating Motor Buses," *Electric Railway Journal* 58 (Aug. 27, 1921): 319.

33. "Where the Bus Fits Into the Picture," *Electric Railway Journal* 64 (Aug. 9, 1924): 189.

34. "Interurbans Co-operate with Independent Bus Lines," *Electric Railway Journal* 63 (Mar. 1, 1924): 329.

35. "Auto-Bus Line in Seattle," *Electric Railway Journal* 47 (Jan. 22, 1916): 164.

36. Charles E. Thompson, "Bus Operation," *Electric Railway Journal* 64 (Sept. 27, 1924): 488.

37. Henry W. Blake and Harry L. Brown, "Abandonments Are Not Always Justified," *Electric Railway Journal* 64 (Dec. 13, 1924): 983.

38. "Railway Operates Passenger Automobiles for Intercity Service," *Electric Railway Journal* 63 (April 12, 1924): 581.

39. "Buses Replace Street Cars," *Bus Transportation* 11 (Aug. 1932): 372.

40. "Bus Operation in Connection with Electric Railways," *Bus Transportation* 6 (Feb. 1927): 93.

41. Harre W. Demoro, *California's Electric Railways* (Glendale, CA: Interurban Press, 1986), 12.

42. "Pacific Electric Railway Experiments with Motor Bus Feeders," *Electric Railway Journal* 43 (Aug. 19, 1916): 314.

43. "Pacific Electric Co-ordinates Rail and Bus Service," *Railway Age,* n.s., 82 (Mar. 26, 1927): 1021.

44. "414 Buses Now Used by Pacific Electric Railway," *Electric Railway Journal* 75 (Aug. 1931): 431.

45. "Express and Freight Service on the Detroit United Lines, *Electric Railway Journal* 43 (May 9, 1914): 1027.

46. "Joint Trolley-Auto Freight Arrangements," *Electric Railway Journal* 42 (Apr. 15, 1916): 754.

47. Charles J. Laney, "Interurbans Have Not Done Duty," *Electric Railway Journal* 52 (Nov. 30, 1918): 954.

48. A. B. Cole, "Electric Railways are in a Position to Haul More Freight," *Electric Railway Journal* 51 (May 11, 1918): 898–99.

49. "Tractors and Trailers Used by North Shore Line," *Railway Age,* n.s., 83 (Oct. 27, 1927): 807–9.

50. T. H. Stoffel, "Freight Service," *Electric Railway Journal* 64 (Sept. 27, 1924): 507.

51. Marlette, *Electric Railroads of Indiana,* 21.

52. "Modernized Freight Service Attracting New Business," *Electric Railway Age* 75 (Mar. 1931): 145

53. Hilton and Due, *Electric Interurban Railways,* 186–87.

54. Ernest Gonzenback, "The Electric Railway and the Automobile," *Electric Railway Journal* 47 (Jan. 18, 1916): 75.

Chapter 7

1. "Rail Auto-Cars," *Railroad Gazette,* n.s., 37 (June 24, 1904): 37.

2. "Traffic Conveyed by Motor Cars," *Railway Age,* n.s., 39 (Aug. 28, 1903): 694.

3. "Gasolene Cars for Interuban Service," *Street Railway Journal* 22 (Sept., 19, 1903): 584.

4. "Some Gasolene Motor Cars for Railroad Use," *Railroad Gazette,* n.s., 40 (1906): 605–6.

5. "Gasolene Motor Cars for Railroad Service," *Railroad Gazette,* n.s., 42 (June 7, 1907): 778.

6. "The Union Pacific Gasoline Motor Car," *Railroad Gazette,* n.s., 43 (Mar. 10, 1905): 200.

7. "Railroad Motor Cars," *Railroad Gazette,* n.s., 39 (July 7, 1905): 2.

8. Bernard Miklejohn, "New Motors on Railroads: Electric and Gasoline Cars Replacing the Steam Locomotive," *World's Work* 13 (June 1907): 8446.

9. Ibid., 8452.

10. "The McKeen Motor Cars," *Railway Age Gazette,* n.s., 46 (Jan. 15, 1909): 111.

11. "McKeen Motor Car for Norfolk & Southern," *Railroad Gazette,* n.s., 47 (Dec. 17, 1909): 1154.

12. John H. White, *The American Passenger Car,* pt. 2 (Baltimore: Johns Hopkins University Press, 1978), 594–95.

13. "The Gasoline Car for Interurban Service," *Railroad Gazette,* n.s., 51 (June 1, 1906): 556.

14. "Stover Twelve-Passenger Motor Car," *Railroad Gazette,* n.s., 53 (Feb. 28, 1908): 292.

15. "The Strang Gasolene-Electric Rail Motor Car," *Railroad Gazette,* n.s., 40 (Feb. 23, 1906): 188.

16. "One-Man Gasoline Motor Car," *Electric Railway Journal* 57 (Feb. 26, 1921): 497.

17. C. W. Foss and C. B. Peck, "Order for Automotive Equipment," *Railway Age,* n.s., 80 (Jan. 2, 1926): 110.

18. "Comparative Cost of Fuel Oil Motor Car and Steam Locomotive Service on Rock Island," *Railway Age Gazette* 54 (Mar. 26, 1909): 699.

19. "Motorized Light Service Rail Transportation," *Railway Age,* n.s., 69 (Apr. 19, 1924): 977, 979.

20. W. G. Besler, "This Country's First Diesel-Electric Locomotive," *Railway Age,* n.s., 101 (Aug. 15, 1936): 256.

21. "Warning to Railroads," *Electric Railway Journal* 47 (Feb. 12, 1916): 328.

22. "Rail Cars and Motor Buses Make Good Showing on New York–New Haven Run," *Highway Transportation* 11 (May 1922): 10.

23. W. L. Bean, "Design Factors of the Gasoline Rail Car," *Railway Age,* n.s., 72 (Sept. 27, 1924): 535.

24. W. R. Stinemetz, "Gas-Electric Cars Cut Railroad Costs," *Electric Railway Journal* 71 (Apr. 7, 1928): 575.

25. C. W. Foss and C. B. Peck, "Order for Automotive Equipment," *Railway Age,* n.s, 80 (Jan. 2, 1926): 109.

26. "New Haven Analyzes Rail Car Operations," *Railway Age,* n.s., 89 (Aug. 9, 1930): 274.

27. "Burlington Gas-Electrics Cut Operating Costs," *Railway Age,* n.s., 90 (Feb. 28, 1931): 439.

28. "Thirty Rail-Motor Cars Ordered Last Year," *Railway Age,* n.s., 92 (Jan. 2, 1932): 68.

29. See Edmund Keilty, *Doodlebug Country: The Rail Motorcar on the Class I Railroads of the United States* (Glendale, CA: Interurban Press, 1982).

30. "FWD High-Speed Rail Bus," *Railway Age,* n.s., 95 (Oct. 21, 1933): 561–62.

31. "Budd-Michelin Car Delivered to the Reading," *Railway Age,* n.s., 93 (Nov. 12, 1932): 669–70.

32. "Pullman Light-Weight Rail Car Designed for High Speeds," *Railway Age,* n.s., 95 (Oct. 7, 1933): 489.

33. "Seaboard Has Many Highway Operations, *Railway Age,* n.s., 100 (Mar. 28, 1936): 546–47.

34. "U.P. Gets Second High-Speed Train," *Railway Age,* n.s., 97 (Oct. 13, 1934): 427–29

35. "Mile-a-Minute Trains Capture Public Imagination," *Railway Age,* n.s., 99 (Nov. 30, 1935): 701–2.

36. Ibid., 701.

37. Mark Reutter, "On the Wings of the Zephyr: The Rise and Fall of America's High-Speed Streamliners, 1934–1960," Indiana Historical Society website, http://www.indianahistory.org/our-services/ (accessed Feb. 23, 2016).

38. Ibid.

39. Ibid.

40. "Mile-a-Minute Trains Capture Public Imagination," 702.

41. "The Burlington Zephyr," *American Experience,* PBS website, http://www.pbs.org/wgbh/americanexperience/features/general-article/streamliners-burlington/, 1 (accessed Aug. 15, 2014).

42. "Illinois Central Lightweight, Alloy-Steel Rail Cars," *Railway Age,* n.s., 109 (Nov. 30, 1940): 819–23.

43. "Is Speed What the Public Wants?" *Railway Age*, n.s., 95 (Aug. 19, 1933): 276.

44. Raymond A. Doster, "1936—A Year of Outstanding Traffic Developments," *Railway Age*, n.s., 102 (Jan. 2, 1937): 16.

45. "I.C.C. Issues Report on Diesel Performance," *Bus Transportation* 15 (Sept. 15, 1936): 434.

46. "High Speeds Demand Good Track," *Railway Age*, n.s., 109 (Nov. 16, 1940): 749-50.

47. Ibid., 781.

Chapter 8

1. Pin-nan Wang, *Rail and Motor Carriers: Competition and Regulation*, published PhD diss. (Philadelphia: University of Pennsylvania, 1932), 16.

2. Ibid., 29.

3. "A Survey of the Bus and Truck Situation," *Railway Age*, n.s., 79 (Dec. 5, 1925): 1023.

4. Ralph Budd, "The Trend in Passenger Travel," *Railway Age*, n.s., 81 (Apr. 28, 1928): 1025-26.

5. W. W. Atterbury, "Looking Ahead in Transportation: Coordination of Trains, Motor Cars, and Airplanes," *American Review of Reviews* 79 (Apr. 1929): 59-61.

6. Carlton Jackson, *Hounds of the Road: A History of the Greyhound Bus Company* (Dubuque, IA: Kendall/Hunt Publishing Co. for Bowling Green University Popular Press, 1984), 12-13.

7. "Motor Cars," *Railway Age Gazette*, n.s., 54 (June 20, 1911); 1586.

8. "A Heavy Highway Car for Railroad Use," *Railway Age*, n.s., 79 (July 4, 1923): 44.

9. J. W. McCloy, "Lower Cost Per Passenger and Greater Reliability," *Electric Railway Journal* 72 (Sept. 8, 1928): 426.

10. Ralph Budd, "Discuss Rail-Highway Co-ordination," *Railway Age*, n.s., 80 (Apr. 24, 1826): 1138.

11. "A Survey of the Bus and Truck Situation," 1025.

12. John C. Emery, "Bus or Train—Or Both?" *American Review of Reviews* 74 (Sept. 1926): 286.

13. For a take on the railroad industry's view of bus evolution, see R. V. Fletcher, *Highway Motor Transportation* ([Washington, DC]: Association of American Railroads, 1945).

14. "Fare Cuts Presage New Bus-Rail War in Texas," *Bus Transportation* 9 (Dec. 1930): 690.

15. "Bus Lines Meet 2-Cent Rate of Railroads," *Bus Transportation* 10 (Mar. 1931): 160.

16. "Railroad Fare Cuts Bring Negative Results," *Bus Transportation* 10 (May 1931): 221.

17. "$40,048,607 Invested by Class I Railroads in Bus and Truck Operation," *Bus Transportation* 9 (Dec. 1930): 662.

18. "ICC Disclaims Right to Rule Railroad Buses," *Bus Transportation* 8 (Apr. 1929): 229.

19. "Railways and Motor Transportation," *Railway Age*, n.s., 79 (Dec. 12, 1926): 1075.

20. "Utility Commissioners Discuss Motor Transport," *Railway Age*, n.s., 83 (Nov. 26, 1927): 1076.

21. "Railways Handicapped in Meeting Highway Competition," *Railway Age*, n.s., 93 (Dec. 3, 1932): 812.

22. Frank R. Fageol, "This Transportation Business," *Railway Age*, n.s., 86 (Apr. 27, 1929): 1023

23. "ICC Rebukes Railroads on Costly Propaganda," *Bus Transportation* 14 (Feb. 1935): 87.

24. "The Highway-Railroad Committee to Study Common Problems of Highway Users and Railroads," *Bus Transportation* 11 (Nov. 1932): 461.

25. "Trend in Railroad Bus Use Increasing," *Bus Transportation* 9 (June 1930): 319–21.

26. Wang, *Rail and Motor Carriers*, 109.

27. F. J. Scarr, "The Proper Place of Trains and Buses," *Railway Age*, n.s., 82 (June 25, 1927): 2052.

28. Robert H. Newcomb, "New Haven Optimistic About Its Bus Operations," *Railway Age*, n.s., 79 (Dec. 5, 1925): 1051.

29. Ibid., 1052.

30. "New Haven Adds 57 Buses to Fleet," *Railway Age*, n.s., 80 (Mar. 27, 1926): 923.

31. "New Haven Bus and Truck Subsidiaries Work Together," *Railway Age*, n.s., 101 (July 25, 1936): 150.

32. "The Pro and Con of Railroad Bus Operation," *Railway Age*, n.s., 81 (Dec. 25, 1926): 1291.

33. "B. & M. Buses Improve Rail Service," *Railway Age*, n.s., 80 (May 22, 1926): 1396.

34. "Bus Experts Discuss Rail-Highway Co-ordination," *Railway Age*, n.s., 82 (June 213, 1927): 2051; Howard F. Fritch, "B. & M. Finds Merit in Co-ordinated Highway Service," *Railway Age*, n.s., 82 (Mar. 26, 1927): 1035.

35. R. C. Boyden, "Boston & Maine's Efficient Maintenance Plan," *Railway Age*, n.s., 88 (Mar. 22, 1930): 715.

36. "Million Dollars Saved by Motor Coach Substitution," *Railway Age*, n.s., 91 (July 25, 1931): 144.

37. "Maine Central Operates Buses and Trucks," *Railway Age*, n.s., 78 (Mar. 27, 1926): 925.

38. George D. Ogden, "Co-ordinated Transportation," *Railway Age*, n.s., 88 (Apr. 26, 1930): 1028.

39. "The T.A.T. Air Rail Service Coast to Coast," *Railway Age*, n.s., 87 (July 6, 1929): 12–16.

40. "Reading Motor Coaches Save 391,000 Train Miles," *Railway Age*, n.s., 90 (June 27, 1931): 1255.

41. M. P. Steinberger, "How the B. & O. Operates Its Unique New York Terminal Service," *Railway Age*, n.s., 82 (Jan. 22, 1927): 342.

42. "When Motor Coaches Fail," *Railway Age*, n.s., 89 (Oct. 25, 1930): 879.

43. "Chicago & Alton Purchases 8-Wheel Highway Coaches," *Railway Age*, n.s., 80 (May 22, 1926): 1400.

44. Emery, "Bus or Train—Or Both?" 107–8.

45. "Notes and News," *Railway Age*, n.s., 27 (Mar. 10, 1899): 166.

46. See "Santa Fe Finds Buses of Value in New Services," *Bus Transportation* 6 (Jan. 1927): 1–4; and "Indian Detours Attract Passengers," *Railway Age*, n.s., 98 (June 22, 1935): 983–84.

47. See "The Santa Fe's Far-Flung Bus System," *Railway Age*, n.s., 103 (Nov. 27, 1937): 765–66.

48. "Burlington Substitutes Motor Coaches for Trains," *Railway Age*, n.s., 90 (Feb. 28, 1931): 459.

49. "Burlington Coaches Reach the Pacific," *Railway Age*, n.s., 98 (Jan. 19, 1935): 85.

50. "In One Year—A 4,000-Mile System," *Railway Age*, n.s., 88 (Feb. 22, 1930): 520.

51. See "Union Pacific Buys Large Nebraska Motor Coach Line," *Railway Age*, n.s., 81 (July 27, 1929): 317.

52. "Five Bus Lines Merge with Union Pacific," *Bus Transportation* 9 (July 1930): 416.

53. "Large Scale Operations Increase Efficiency," *Railway Age*, n.s., 81 (Nov. 27, 1926): 1069–70.

54. "Southern Pacific Involved in Motor Coach Merger," *Railway Age*, n.s., 86 (June 22, 1929): 1503.

55. Ibid.

56. H. Roger Grant, *Getting Around: Exploring Transportation History* (Malabar, FL: Krieger Publishing, 2003), 38.

57. John A. Jakle and Keith A. Sculle, *Motoring: The Highway Experience in America* (Athens: University of Georgia Press, 2008), 189.

58. "Looking Over the Biggest Bus System," Railway Age, n.s., 82 (Jan. 22, 1927): 329.

59. "The Pickwick Motor Coach Sleeper," *Railway Age,* n.s., 85 (Sept. 22, 1928): 587.

60. For a historical overview of the Greyhound story, see Margaret Walsh, *Making Connections: The Long-Distance Bus Industry in the USA* (Aldershot, UK: Ashgate, 2000).

61. "Historical Timeline," Greyhound website, https://services.greyhound.com/en/about /historicaltimeline.aspx. See also Margaret Walsh, "Tracing the Hound: The Minnesota Roots of the Greyhound Bus Corporation," *Minnesota History* 49 (Winter 1985): 310–21.

62. "Major Cross-Country Bus Systems Form $30,000,000 Merger," *Bus Transportation* 8 (June 1929): 350.

63. "Transcontinental Motor Coach Service Tested," *Railway Age,* n.s., 86 (Feb. 9, 1929): 358.

64. S. R. Sundstrom, "The Story of Pennsylvania Greyhound Lines," *Railway Age,* n.s., 95 (Oct. 21, 1933): 566.

65. "Greyhound Assumes Dominant Place in New England Field," *Bus Transportation* 18 (Feb. 1939): 100.

66. "Make-Up of the Greyhound Lines," *Railway Age,* n.s., 103 (Dec. 26, 1937): 917, 919.

67. "Rail-Highway Transportation Takes New Spurt," *Railway Age,* n.s., 100 (Mar. 28, 1936): 553.

68. H. G. Murphy, "How Burlington is Co-ordinating Rail and Highway Service," *Railway Age,* n.s., 106 (May 27, 1939): 913.

69. Earle S. Montgomery, "The Story of the New England Transportation Co.," *Railway Age,* n.s., 87 (Aug. 24, 1929): 494.

70. "New York's Newest Terminal," *Railway Age,* n.s., 88 (Feb. 22, 1930): 515.

71. "Railroads Operating More Than 5,200 Buses," *Bus Transportation* 18 (Aug. 1939): 433.

Chapter 9

1. "Influence of the Motor Truck," *Freight: The Shippers' Forum* 12 (Apr. 1911): 117.

2. "The Motor Truck a Strong Link in the Transportation Chain," *Freight: The Shippers' Forum* 12 (Jan. 1911): 14.

3. "Motor Trucks Developed from the Locomotive," *Freight: The Shippers' Forum* 1 (Mar. 1911): 79.

4. "Motor Trucks—The New Freighters," *World's Work* 23 (Jan. 1912): 282.

5. "Highway Transport Committee Organized," *Railway Age Gazette* 63 (Nov. 16, 1917): 891.

6. "Adaptability," *Literary Digest* 55 (Sept. 23, 1917): 60.

7. Harry Wilkin Perry, "National Crisis Brings Motor-Trucks to the Fore," *Highway Transportation* 8 (Dec. 1918): 8.

8. "Influence of the Motor Truck," *Freight: The Shippers' Forum* 12 (Apr. 1911): 117.

9. "The High Cost of Horses," *Highway Transportation* 8 (July 1918): 19.

10. P. W. Litchfield, "History's Lesson to the Motor Truck," *Highway Transportation* 8 (Nov. 1918): 6–7, 24.

11. Major Elihu Church, "Motor Transportation of the Future," *Highway Transportation* 15 (Aug. 1925): 8.

12. "How War Hit Auto Industry," *Highway Transportation* 8 (May 1919): 19.

13. "Motor Truck Great Factor in the War," *Highway Transportation* 8 (May 1919): 42.

14. John Martin, "The Importance of Stabilizing the Highway Transportation Industry," *Highway Transportation* 8 (June 1919): 21.

15. "The Trucks Aid Railroads," *Highway Transportation* 11 (Feb. 1922): 7.

16. R. E. Fulton, "How the Problem of Motor Transportation Will Be Solved," *Highway Transportation* 8 (May 1919): 10–13.

17. Donald McLeod Lay, "Trailers Increase Efficiency of Motor Trucks," *Highway Transportation* 10 (Oct. 1920): 9.

18. "Motor Trucks Come to the Aid of the Railroads," *Highway Transportation* 10 (May 1921): 16.

19. Elisha Lee, "Co-ordination of Motor Transport and Railroads," *Railway Age,* n.s., 74 (Feb. 10, 1923): 386.

20. George M. Graham, "Relation of Highway Transportation to Increased Production," *Highway Transportation* 10 (Aug. 1920): 33–34.

21. "Motor Trucking Not Yet a Science," *Electric Highway Journal* 57 (Apr. 16, 1921): 710–11.

22. "Relation of Motor Transport to Other Transport Agencies," *Railway Age,* n.s., 75 (Nov. 24, 1923): 965.

23. Robert C. Wright, "The Freight Car Yields to the Truck," *Highway Transportation* 14 (Aug. 1924): 9.

24. Church, "Motor Transportation of the Future," 25.

25. T. R. Dahl, "The Correlation of Transportation Facilities," *Good Roads Magazine* 69 (Feb. 1926): 68. For details of several states regarding taxes paid by buses and trucks, see "The Taxes Buses and Trucks Pay," *Railway Age,* n.s., 82 (Apr. 23, 1927): 1291–95.

26. Dahl, "The Correlation of Transportation Facilities," 67.

27. "Trucks, Tractors and Trailers Widely Used," *Railway Age,* n.s., 82 (June 25, 1927): 2039.

28. *Facts And Figures of the Automobile Industry* (New York: National Automobile Chamber of Commerce, 1928), 6, 33.

29. A. D. Ferguson, "An Analysis of Motor Truck Taxation," *Railway Age,* n.s., 88 (Jan 25, 1930): 293.

30. "Unwarranted Government Expenditures—A Highway Example," *Railway Age,* n.s., 78 (Mar. 11, 1933): 357.

31. Ibid., 355.

32. Ibid.

33. Harold L. Whitridge, *The I.C.C. vs. Class I Railroads* (New York: Simons-Boardman Publishing, 1931), 35.

34. "Motor Manufacturers and Highway Transportation Policy," *Railway Age,* n.s., 93 (Oct 8, 1932): 494.

35. "Survey of Truck Transportation," *Railway Age,* n.s., 772 (Sept. 10, 1932): 363.

36. "Room Enough for Both," *Saturday Evening Post* 205 (May 27, 1933): 20.

37. *Official Motor Freight Guide, Cartage Section, Warehouse Section,* 2 (Chicago: Universal Guide Corporation, 1935), 16.

38. F. Leslie Jacobus, ed., *The Motor Truck Red Book,* 1936 ed. (New York: Traffic Publishing, Sept. 1935), 10.

39. *Motor Truck Facts,* 1940 ed. (Detroit: Automobile Manufacturers Association, [1940]), [3].

40. Oliver C. Miller, "Cutting Costs With Diesel-Powered Trucks," *Official Motor Freight Guide* 1 (Nov. 1934): 10–11.

41. *Report of Subcommittee on Motor of Railroad Committee for the Study of Transportation* (N.p.: Association of American Railroads, Aug. 1945).

42. "Motor Trucks and Local Deliveries," *Freight: The Shippers' Forum* 12 (Nov. 1911): 297.

43. "The Motor Truck a Strong Link in the Transportation Chain," *Freight: The Shippers' Forum* 12 (Jan. 1911): 14.

44. W. M. Oestreicher, "Local Freight and the Motor Haul," *Freight: The Shippers' Forum* 11 (Nov. 1910): 329–30.

45. "Motor Truck Use for Short Hauls," *Highway Transportation* 8 (Apr. 1919): 17. See also "Cost of Freight by Motor Truck," *Railway Age,* n.s., 67 (July 18, 1919): 112.

46. "Advantages of Motor Trucking," *Trade and Transportation* 14 (June 1913): 14–15; "Motor Truck Operating Cost," *Trade and Transportation* 14 (Oct. 1913, 19.

47. William Rice Smith, "Motor Truck Traffic," *Good Roads Magazine* 49 (Jan. 8, 1916): 28.

48. "Store-Door Delivery in New York City," *Railway Age,* n.s., 64 (Feb. 1, 1918): 241.

49. "Auto Truck Line Between Waukegan and Chicago," *Railway Age,* n.s., 64 (Mar. 1, 1918): 476.

50. F. Van Z. Lane, "Motor Truck Fits into Short Haul Problem from Every Economic Standpoint," *Freight Transportation Digest* 1 (July 16, 1919): 6, 7.

51. "How The Motor Truck Fits into The Cotton Industry—Plantation and Mill," *Freight Transportation Digest* 1 (Nov. 1919): 3, 5.

52. "Flour Mill Finds Motor Truck Cheaper Than L.C.L. Rates," *Freight Transportation Digest* 1 (July 16, 1919): 14.

53. "Boss of Packing House Garage No Longer Laughs at Charts and Figures," *Freight Transportation Digest* 2 (Aug. 1920): 6.

54. Arthur Capper, "Mid-West Farmers Waiting for Trucks," *Highway Transportation* 9 (June 1920): 48.

55. "More Cattle Moved by Truck," *Railway Age,* n.s., 75 (Mar. 1, 1930): 543.

56. "Freight by Automobile," *Railway Age,* n.s., 68 (May 7, 1920): 1371.

57. "Highway Freight Terminals Needed," *Highway Transportation* 10 (Dec. 1920): 20.

58. "Motor Truck Haulage in Chicago," *Electric Railway Journal* 58 (July 23, 1921): 135.

59. "Detroit Creamery Hauls More Milk by Truck Than by Rail," *Freight Transportation Digest* 2 (Mar. 1920): 3–4.

60. Francis W. Davis, "Motor Truck Transportation," *Railway Age,* n.s., 69 (Dec. 10, 1920): 1012–13.

61. "Motor Cars in Trap Car Service," *Railway Age,* n.s., 69 (Dec. 17, 1920): 1049.

62. *Freight Transportation Digest* 3 (Apr. 1921): 2.

63. G. Marks, "Modern Methods of Handling Package Freight," *Railway Age,* n.s., 72 (Feb 25, 1922): 469.

64. "Meeting Motor Truck Competition," *Railway Age,* n.s., 71 (Sept. 3, 1921): 432.

65. Talcott Williams, "The Railroads—National Highways," *Independent* 102 (June 5, 1920): 320.

66. "The Railroads' Ruinous Rivals," *Literary Digest* 59 (June 4, 1921): 15.

67. "Motor Trucks for Handling Company Materials," *Railway Age*, n.s., 69 (Dec. 24, 1926): 1097.

68. H. C. Hitchborn, "Railroad Livestock Traffic Dwindles," *Railway Age*, n.s., 92 (June 25, 1932): 1056–58.

69. C. V. Beck, "Trucks Making Heavy Inroads into Short Haul Coal Traffic," *Railway Age*, n.s., 92 (Feb. 6, 1932): 254–57.

70. "What Does It Cost to Operate Buses and Trucks?" *Railway Age*, n.s., 80 (Mar. 27, 1925): 920.

71. L. B. Young, "Store-Door Service Proves Successful," *Railway Age*, n.s., 87 (Sept. 28, 1929): 775, 777.

72. "Down-to-Earth Facts on Produce Trucking," *Railway Age*, n.s., 109 (Oct. 12, 1940): 515.

73. "What About the 'Gypsy' Trucker?" *Railway Age*, n.s., 106 (Jan. 21, 1939): 160–61.

74. "Handling L.C.L. Freight in Interchange by Motor Trucks," *Railway Age*, n.s., 69 (Dec. 10, 1920): 107.

75. "Less-Car-Load Shippers Co-operate in Car Loading," *Railway Age*, n.s., 69 (Dec. 31, 1920): 1169–70.

76. "Handling L.C.L. Freight by Tractor and Trailer Trucks," *Railway Age Gazette* 62(Apr. 13, 1917): 792–93.

77. "Expediting the Movement of L.C.L. Freight," *Railway Age*, n.s., 69 (Aug. 5, 1920): 219.

78. "The Trucks Aid Railroads," *Highway Transportation* 11 (Feb. 1922): 26, 28.

79. Ibid., 7.

80. "The L.C.L. Problem and Motor Trucks," *Railway Age*, n.s., 73 (Nov. 25, 1922): 970.

81. R. D. Sanger, "Delivering Less Than Carload Freight by Motor Truck," *Highway Transportation* 14 (Oct. 1924): 5.

82. "Recovering L.C.L. Traffic by Motor Truck Operation," *Railway Age*, n.s., 85 (Sept. 22, 1928): 576; "L.C.L. Traffic Recovery by Motor Truck Operation," *Railway Age*, n.s., 85 (Sept. 22, 1928): 580.

83. "The Trucks Aid Railroads," 30.

84. "Long Distance Automobile Freight Trip," *Railway Age Gazette* 50 (May 12, 1911): 1122.

85. "Moreland Auto-Rail Truck," *Railway Age Gazette* 63 (Dec. 28, 1917): 1188.

86. Avery Turner, Letter to the Editor, "Motor Freight Trucks on Highways," *Railway Age Gazette* 62 (Dec. 14, 1917): 1067.

87. "Trucks Coming into Their Own," *Highway Transportation* 8 (July 1918): 15.

88. "Thirty Army Trucks on 600 Mile Run," *Railway Age Gazette* 63 (Dec. 21, 1917): 1147.

89. "Freight by Automobile from Detroit to Newport News," *Railway Age Gazette* 63 (Dec. 28, 1917): 1190.

90. "Army Motors Detroit to Seaboard," *Railway Age*, n.s., 63 (Jan. 11, 1918): 140.

91. "The New York and Connecticut Fright Line," *Railway Age*, n.s., 64 (Mar. 22, 1918): 731.

92. "Parcel Post Routes by Motor Trucks to Be Extended," *Railway Age*, n.s., 64 (Jan. 11, 1918): 104.

93. "Trucks Coming into Their Own," 15.

94. "Motor Truck Lines as Business Aids," *Highway Transportation* 8 (Aug. 1918): 16–17.

95. "Seven Hundred Automobile Stock-cars," *Railway Age*, n.s., 67 (Dec. 26, 1919): 1262.

96. "Long Distance Overland Haulage by Highway Freight Trains," *Highway Transportation* 11 (Oct. 1921): 23.

97. Arthur H. Blanchard, "The Relation of Highway Transport Service to Economic Highway Improvement," *Highway Transportation* 11 (Mar. 1921): 15.

98. "Long-Distance Freight by Automobile," *Railway Age,* n.s., 68 (May 21, 1920): 1437.

99. "A Plea for the Horse Truck," *Railway Age,* n.s., 72 (Jan. 21, 1922): 246.

100. William M. Jardine, "Motor Trucks Do Not Destroy Highways," *Highway Transportation* 15 (Nov. 1925): 9.

101. L. B. Young, "Truck Competition Is Taking the Railways' Carload Traffic," *Railway Age,* n.s., 89 (Sept. 27, 1930): 649.

102. "Why Truck Lines Attract Traffic," *Railway Age,* n.s., 75 (Mar. 22, 1930): 729.

103. John C. Emery, "Wider Use of Motor Transport," *Railway Age,* n.s., 70 (Jan. 3, 1931): 107–8.

104. George P. McCallum, "Is It Worth It?" *Bus Transportation* 11 (Dec. 1932): 493–94; "Highway Users Present a United Front," *Bus Transportation* 11 (Nov. 1932): 471.

105. "60-Hour Week for Truck Drivers," *Railway Age,* n.s., 105 (July 30, 1938): 188–89.

106. "Says Motor Drivers Can Stand 60-Hr. Week," *Railway Age,* n.s., 82 (July 24, 1937): 108, 111.

107. "Motor Truck Service in Competition with a Railway," *Railway Age Gazette* 53 (Sept. 20, 1912): 510–11.

108. "Electric Motor Trucks for New York Railways," *Electric Railway Journal* 44 (Sept. 5, 1914): 427.

109. "Motor Truck Service for Stores Department," *Electric Railway Journal* 48 (Nov. 18, 1916): 1068–69.

110. "Motor Trucks Replace Trap Cars," *Railway Age Gazette* 63 (Sept. 7, 1917): 427.

111. "Committee to Promote Use of Motor Truck," *Railway Age Gazette* 63 (Nov. 23, 1917): 960.

112. "Why Railroads Like Motor-Trucks," *Literary Digest* 56 (Dec. 28, 1918): 85. See also "Freight-Cars Vs. Motor-Trucks," *Literary Digest,* 55 (Nov. 3, 1917): 21.

113. B. F. Fitch, "Motorizing Railroad Terminals," *Scientific American* 176 (May 1920): 448–50.

114. "The Theory and Practice of Store Door Delivery," *Railway Age,* n.s., 72: 13 (Apr. 1, 1922): 825–28.

115. C. S. Mott, "Railroads Recognize Truck Utility," *Highway Transportation* 11 (Feb. 1922): 18–19, 22.

116. R. S. Parsons, "What the Truck Means to the Railroad," *Highway Transportation* 12 (Aug. 1922): 27.

117. Elisha Lee, "The Relation of the Railroad and Motor Vehicle," *Highway Transportation* 12 (Feb. 1923): 15.

118. *Facts about Railroad Use of Motor Trucks* (New York: National Automobile Chamber of Commerce, c. 1924), 3. See also "Meeting the Competition of Buses and Trucks," *Railway Age,* n.s., 70 (Dec. 12, 1925): 1077–82.

119. "Railways and Motor Transportation," *Railway Age,* n.s., 79 (Dec. 12, 1924): 1076.

120. "Superintendents Favor Motor Vehicle as Railroad Auxiliary," *Railway Age,* n.s., 83 (July 23, 1927): 174.

121. Robert C. Wright, *The Freight Car Yields to the Truck* (New York: National Automobile Chamber of Commerce, 1924), 9.

122. "Railway Plus Truck," *Highway Transportation* 15 (Apr. 1926): 12 and 19.

123. "Bus and Truck Operation Should Be Co-ordinated," *Railway Age,* n.s., 80 (Jan. 16, 1926): 231–33; "Significant Changes In Railroad Freight Traffic," *Railway Age,* n.s., 85 (July 21, 1928): 96–97; A. J. Brosseau, "The Motor Truck: A Helper, Not a Competitor, of the Railways," *Railway Age,* n.s., 87 (July 27, 1929): 297.

124. "Railways in 1929 Ordered 1,764 Trucks, 390 Motor Coaches," *Railway Age,* n.s., 87 (Dec. 28, 1929): 1535.

125. Ping Nan Wang, *Rail and Motor Carriers: Competition and Regulation,* published PhD diss. (Philadelphia: University of Pennsylvania, 1932); William J. Cunningham, "Correlation of Rail and Highway Transportation," *American Economic Review* 24 (Mar. 1934): 47–56.

126. Wang, *Rail and Motor Carriers,* 99, 101.

127. Ibid., 6.

128. Cunningham, "Correlation of Rail and Highway Transportation," 52.

129. "Rail and Motor Rivals Getting Together," *Literary Digest* 115 (Feb. 18, 1933): 40.

130. "Truck-Rail Co-ordination Suggested by Eastman," *Bus Transportation* 13 (Apr. 1934): 146–47.

131. "What Wage for Transport Labor?" *Railway Age,* n.s., 97 (Dec. 1, 1934): 734–35.

132. "Other Expenditures in Prospect," *Railway Age,* n.s., 97: 22 (Dec. 1, 1934): 731–33.

133. John R. Turney, "Railway and Truck, Co-partners," *Railway Age,* n.s., 99 (Nov. 23, 1935): 679.

134. Charles Layng, "R. R. Motor Vehicle Use Expands," *Railway Age,* n.s., 108 (Jan. 6, 1940): 101. See also "Katy Begins Co-Ordinated Service," *Railway Age,* n.s., 108 (Jan. 27, 1940): 222–23.

135. "Making Rates to Move the Freight," *Railway Age,* n.s., 108 (May 25, 1940): 936–38.

136. "Take Strings off RR Truck Use," *Railway Age,* n.s., 108 (June 22, 1940): 1135–36.

137. "Motor Transportation," *World's Work* 43 (Dec. 1921): 126. See also "Boston & Maine Competes with Motor Trucks," *Railway Age,* n.s., 71 (Sept. 2, 1921): 467.

138. C. L. Bardo, "The Railroads and Highway Transport," *Railway Age,* n.s., 77 (Dec. 13, 1924): 1069.

139. "Erie Adopts Direct Freight Delivery at New York," *Railway Age,* n.s., 72 (Jan. 21, 1922): 233.

140. "Boston & Main Reorganization Proposed," *Railway Age,* n.s., 77 (Dec 27, 1924): 1157–58.

141. F. J. Carey, "Seven Years of Trucking," *Railway Age,* n.s., 92 (Feb. 27, 1932): 369. See also "Analyze Results of Pick-Up and Delivery Service," *Railway Age,* n.s., 95 (July 22, 1933): 160–62.

142. "Lehigh Valley Substitutes Trucks for Trains and Switching," *Railway Age,* n.s., 80 (Feb. 27, 1926): 543–47; "Long Island Uses Motor Vehicles to Reduce Congestion," *Railway Age,* n.s., 81 (July 24, 1926): 157–59; "B. & O. Establishes Constructive Station at New York," *Railway Age,* n.s., 82 (May 28, 1927): 1665–66; "Trucks Handle Freight on B. & O.," *Railway Age,* n.s., 83 (Aug. 27, 1927): 407–8.

143. For the Erie Railroad, see "Motor Truck-Inland Station Plan Speeds Erie Service at New York," *Railway Age,* n.s., 79 (Nov. 24, 1934): 651–52. For the Pennsylvania Railroad, see, for example, "Pick-Up and Delivery Service Proving Worth" *Railway Age,* n.s., 96 (May 26, 1934): 785–87.

144. George M. Crowson, "Economic Advance of South Sped by Railroad Development," *Railway Age,* n.s., 82 (May 21, 1927): 1533.

145. "Store-Door Service Adopted by Southeastern Roads," *Railway Age,* n.s., 78 (Apr. 21, 1933): 596.

146. Loomis L. Kelly, "Norfolk & Western Builds Freight Station at Bluefield," *Railway Age,* n.s., 79 (Oct. 10, 1925): 663–66.

147. "Store-Door Service Adopted by Southeastern Roads," 595.

148. John C. Emery, "Railways Extend Highway Service," *Railway Age,* n.s., 96 (Jan. 27, 1934): 162.

149. "St. Louis Roads Coordinate Rail and Motor Service," *Railway Age,* n.s., 74 (June 30, 1923): 1690, 1691.

150. "Wabash Uses Tractors and Trailers," *Railway Age,* 82 (Apr. 23, 1927): 1303.

151. "Rock Island Finds Tractors and Trailers Valuable," *Railway Age,* n.s., 86 (Jan. 26, 1929): 285.

152. "Rock Island 'Ferry Service' for Truck Trailers Successful," *Railway Age,* n.s., 98 (Jan. 19, 1935): 86.

153. "Keeshin Controls Many Companies," *Railway Age,* n.s., 82 (Oct. 24, 1936): 603.

154. Carl R. Gray, "One Railroad's Answer to the Truck Question," *Railway Age,* n.s., 89 (July 26, 1930): 201.

155. "Illinois Central Goes to Trucks," *Railway Age,* n.s., 107 (Sept. 23, 1939): 447.

156. "Trucks, Rails Find Freight War in Mountain States Is Costly," *Business Week* 89 (May 6, 1931): 30.

157. Ibid., 32.

158. "Southwestern Rails Battle Trucks with Free Store-Door Service," *Business Week* 105 (Sept. 9, 1931): 15.

159. Charles Layng, ""Railway Motor Transport Marches On," *Railway Age,* n.s., 106 (Jan. 7, 1939): 93-94. Within Texas, see "S.P. Trucks in Texas," *Railway Age,* n.s., 107 (July 15, 1939): 117-18.

160. Layng, "Railway Motor Transport Marches On," 93.

161. "St. Louis Southwestern," *Railway Age,* n.s., 69 (Dec. 3, 1920): 959.

162. "Operating a Co-ordinated Rail-Motor Truck Service," *Railway Age,* n.s., 83 (Apr. 26, 1930): 1014.

163. "Store-Door Service—Can the Railroads Afford It?" *Railway Age,* n.s., 94 (Mar. 25, 1933): 453-54.

164. "Speed and Rail-Highway Co-ordination Bring Merchandise to Southern Pacific," *Railway Age,* n.s., 100 (Jan. 25, 1936): 182.

165. Ibid., 183-84.

166. H. C. Murphy, "How Burlington is Co-Ordinating Rail and Highway Service," *Railway Age,* n.s., 106 (May 27, 1939): 913.

167. "Selling Railroad Freight Service," *Railway Age,* n.s., 107 (Nov. 25, 1939): 826-27.

168. "Union Pacific a Storedoor Pioneer," *Railway Age,* n.s., 97 (Dec. 22, 1934): 834.

169. For the Texas & Pacific Motor Transport Company subsidiary of the Texas & Pacific Railway, see "Winning Traffic Back from the Truck Lines," *Railway Age,* n.s., 88 (May 24, 1930): 1270-73; and for the Columbia Terminals Company subsidiary of the Missouri Pacific Railroad in Kansas, Missouri, Nebraska, and Louisiana, see "Trucks Give Flexible Operation," *Railway Age,* n.s., 107 (Aug. 19, 1939): 290.

170. "A Century of Express Service," *Railway Age,* n.s., 106 (Mar. 4, 1939): 365-66.

171. "Our Largest Truck Operator—The Express Agency," *Railway Age,* n.s., 88 (Jan. 25, 1930): 278.

172. "Railway Express Agency Begins Highway Freight Service," *Railway Age,* n.s., 9 (Aug. 27, 1932): 296-99.

173. Irvin Foos, "Overheard in Washington," *Bus Transportation* 13 (Jan. 1934): 13.

174. "Fourteen Thousand Trucks Comprise Express Fleet," *Railway Age*, n.s., 108 (Jan. 27, 1940): 215.

175. "Our Largest Truck Operator—The Express Agency," 281.

176. Roy Stratton, "'Container Car' Co-ordinates Steam and Electric Railway and Motor Truck Transportation," *Highway Transportation* 10 (June 1921): 7–8.

177. "The Container System of Freight Transportation," *Railway Age*, n.s., 69 (Sept. 24, 1920): 515, 517.

178. B. F. Fitch, "Employing Motor Trucks in Trap Car Service," *Railway Age*, 68 (Mar. 5, 1920): 681–83; Walter C. Sanders, "Meeting of Railroad Division of the A.S.M.E.," *Railway Age*, n.s., 71 (Dec. 10, 1921): 1147–49; "Containers the Ultimate Answer," *Electric Railway Journal* 57: 5 (Jan. 29, 1921): 219–20; "Container System Creates Freight Service," *Railway Age*, n.s., 72 (Feb. 25, 1922): 475–76.

179. " 'Freightainers' on B. & M. Furnish Door-to-Door Service," *Railway Age*, n.s., 83 (Sept. 24, 1927): 617.

180. For example, see "L. C. L. Transfer by Motor Truck," *Railway Age*, n.s., 86 (Apr. 27, 1929): 1008–12.

181. "Hearing on Container Service," *Railway Age*, n.s., 86 (Feb. 16, 1929): 419.

182. "Rail-Truck Experiment Hangs on I.C.C. Decision," *Business Week* (Dec. 30, 1931): 7–8.

183. David J. DeBoer, *Piggyback and Containers: A History of Rail Intermodal on America's Steel Highway* (San Marino, California: Golden West Books, 1992), 15.

184. "Moving Highway Trucks On Trains," *Railway Age*, n.s., 95 (Nov. 4, 1950): 50–51. Given so little information in print about piggybacks, the experience of one writer aboard a professionally driven truck deserves attention; see Evan McLeod Wylie, "By Piggyback Across America," *American Magazine* 161 (Feb. 1956): 89–90.

185. Henry E. Riggs, "Highway, Railway and Water Transportation," *Highway Transportation* 11 (Feb, 1922): 37.

186. "Early Adoption of Bus and Truck Regulation Indicated," *Railway Age*, n.s., 80 (Jan. 9, 1926): 193–94.

187. "Motor Vehicles Should Pay for Maintaining Highways," *Railway Age*, n.s., 74 (Nov. 24, 1923): 966–67.

188. "New Laws, New Rules," *Railway Age*, n.s., 87 (Nov. 23, 1929): 1251–53.

189. Ibid., 1251.

190. "Mistaken Railway Arguments Against Highway 'Carriers,' " *Roads and Streets* 75 (Oct. 1932): 435.

191. J. J. Pelley, "How Regulate Motor Carriers?" *Railway Age*, n.s., 92 (Feb. 20, 1932): 327.

192. "Show Evils of Unregulated Trucking," *Railway Age*, n.s., 93 (Nov. 5, 1932): 639–42.

193. "Trains vs. Trucks," *Business Week* (Dec. 14, 1932): 12–14.

194. "N.R.A. Regulations Proposed For Railroad Competitors," *Railway Age*, n.s., 95 (Nov. 18, 1933): 727.

195. "History of the trucking industry in the United States," *Wikipedia*, https://en.wikipedia .org/wiki/History_of_the_trucking_industry_in_the_United_States (accessed Jan. 20, 2014).

196. S. W. Fairweather, "Is Truck Regulation a Failure?" *Railway Age*, n.s., 102 (Jan. 8, 1937): 122.

197. "Is Truck Regulation a Failure?" *Railway Age,* n.s., 109 (July 20, 1940): 99.

198. "The Transportation Act Of 1940," Transportation Research International Database, http://trid.trb.org/view.aspx?id=663004 (accessed Aug. 28, 2014).

Conclusion

1. Mark H. Rose, Bruce E. Seely, and Paul H. Barrett, *The Best Transportation System in the World: Railroads, Trucks, Airlines, and American Public Policy in the Twentieth Century* (Columbus: Ohio State University Press, 2004), 49.

2. Margaret Walsh, *The Long Distance Bus Industry in the USA* (Aldershot, UK: Ashgate, 2000), 25.

3. "I.C.C. Opens Hearings on Rail-Highway Service," *Railway Age,* n.s., 89 (Nov. 22, 1930): 1007.

4. "Motor Transport Investigation," *Railway Age,* n.s., 90 (Mar. 14, 1931): 548.

5. Ibid., 549.

6. "Regulation of Water and Motor Carriers Recommended," *Railway Age,* n.s., 94 (Mar. 17, 1934): 377.

7. "A Moderate Increase in Rates," *Railway Age,* n.s., 97 (Dec. 1, 1934): 701–2.

8. "Political Maneuvering and the Railroad Situation," *Railway Age,* n.s., 98 (June 1, 1935): 841.

9. "Motor Carrier Bill Passed," *Railway Age,* n.s., 99 (Aug. 10, 1935): 187.

10. "The 'New Deal' and the Railroads," *Railway Age,* n.s., 99 (Aug. 31, 1935): 265.

11. Fitzgerald Hall, "Transportation Versus Politics," *Railway Age,* n.s., 99 (Oct. 12, 1935): 467.

12. Harold G. Moulton, "The Need for Companies Providing All Forms of Transportation," *Railway Age,* n.s., 99 (Sept. 7, 1935): 305.

13. "Special Motor Vehicle Tax Levies Top State and County Road Costs," *Motor Truck News* 28 (Sept. 1938): 16.

14. "ICC Bureau Head Vetoes R.R's Buying Truck Lines," *Railway Age,* n.s., 100 (April 25, 1936): 698.

15. "Truck Acquisition Not Approved," *Railway Age,* n.s., 101 (Oct. 17, 1936): 565.

16. "Existing Bus Line Refused Permit," *Railway Age,* n.s.,100 (May 23, 1936): 843.

17. For post-1940 assessments of the Motor Carrier Act of 1935, see Dorothy Robin, *Braking the Special Interests: Truck Deregulation and the Politics of Policy Reform* (Chicago: University of Chicago Press, 1987); and Richard D. Stone, *The Interstate Commerce Commission and the Railroad Industry* (New York: Praeger, 1991).

18. James C. Nelson, "The Motor Carrier Act of 1935," *Journal of Political Economy* 44 (Feb.–Dec. 1936): 464–504. See also James C. Nelson, "The Extent and Status of Motor Transport Regulation in the United States," PhD diss., University of Virginia, 1934.

19. Nelson, "The Motor Carrier Act of 1935," 466.

20. Julius H. Parmelee, *The Modern Railway* (New York: Longmans, Green, 1940), pp. 33637.

21. Joseph B. Eastman, "The Motor Carrier and the Transportation Problem," *Bus Transportation* 13 (Oct. 1934): 359–60.

22. Robert E. Gallamore and John R. Meyer, *American Railroads: Decline and Renaissance in the Twentieth Century* (Cambridge, MA: Harvard University Press, 2014), 12.

23. Eastman, "The Motor Carrier and the Transportation Problem," 361.

24. Parmelee, *The Modern Railway,* 494–95.

25. *United States Statistical Abstract* (Washington, DC: Department of Commerce, 1949), 521.

26. "Motor Vehicle Registrations," US Census 2000 on Allcountries.org, http://www.allcountries .org/uscensus/1027_motor_vehicle_registrations.html.

27. *United States Statistical Abstract,* 521.

28. http://www.imua.org/Files/reports/Railroad%20Stock.html.

29. *United States Statistical Abstract,* 542.

30. Books on "Roadside America" coauthored by John A. Jakle and Keith A. Sculle include *The Gas Station in America* (Baltimore: Johns Hopkins University Press, 1994), *Fast Food: Roadside Restaurants in the Automobile Age* (Baltimore: Johns Hopkins University Press, 2002), *Lots of Parking: Land Use in a Car Culture* (Charlottesville: University of Virginia Press, 2004), *Signs in America's Auto Age: Signatures of Landscape and Place* (Iowa City: University of Iowa Press, 2004), *Motoring: The Highway Experience in America* (Athens: University of Georgia Press, 2004), *America's Main Street Hotels: Transiency and Community in the Early Auto Age* (Knoxville: University of Tennessee Press, 2009), and The Garage: *Automobility and Building Innovation in America's Early Auto Age* (Knoxville: University of Tennessee Press, 2013). Coauthored by John A. Jakle, Keith A. Sculle, and Jefferson S. Rogers is *The Motel in America* (Baltimore: John's Hopkins University Press, 1996). Authored by John A. Jakle is *City Lights: Illuminating the American Night* (Baltimore: John's Hopkins University Press, 2001).

31. See John A. Jakle and Keith A. Sculle, *Remembering Roadside America: Preserving the Recent Past as Landscape and Place* (Knoxville: University of Tennessee Press, 2011).

INDEX

Pages numbered in **boldface** refer to illustrations.

DATE DUE

AUG 2 5 2017			
			PRINTED IN U.S.A.